PRACTICING ETHNOGRAPHY IN LAW

Practicing Ethnography in Law

New Dialogues, Enduring Methods

Edited by
June Starr and Mark Goodale

Chapter 5 reprinted from *Studies in Law, Politics & Society*, Vol. 19, Kidder, "Exploring legal culture in law-avoidance societies" (2000), with permission from Elsevier Science.

First published 2002 by
PALGRAVE MACMILLAN™
175 Fifth Avenue, New York, N.Y. 10010 and
Houndmills, Basingstoke, Hampshire, England RG21 6XS
Companies and representatives throughout the world

PALGRAVE MACMILLAN is the global academic imprint of the Palgrave Macmillan division of St. Martin's Press, LLC and of Palgrave Macmillan Ltd. Macmillan® is a registered trademark in the United States, United Kingdom and other countries. Palgrave is a registered trademark in the European Union and other countries.

ISBN 1–4039–6069–0 hardback

Library of Congress Cataloging-in-Publication Data
Practicing ethnography in law: new dialogues, enduring methods/edited by June Starr and Mark Goodale
 p. cm.
 Includes bibliographical references and index.
 ISBN 1–4039–6069–0 hardback
  1. Ethnological jurisprudence—Congresses. I. Starr, June. II. Goodale, Mark.

K190.P73 2002
240'.115—dc21   2002032175

A catalogue record for this book is available from the British Library.

Design by Newgen Imaging Systems (P) Ltd., Chennai, India.

First edition: December, 2002
10 9 8 7 6 5 4 3 2 1

Printed in the United States of America.

In Memory of Our Colleague and Friend
June Starr

CONTENTS

Part II   Reflections on Ethnography in Law

ACKNOWLEDGMENTS

When June Starr asked me to contribute a chapter to this volume in the summer of 1998, I was a graduate student in anthropology preparing to return to Bolivia to finish my doctoral research. I was at once honored and humbled to have been asked to join a project that included many of the scholars with whom I was at that very moment engaged in spirited debate in the pages of grant applications, in drafts of journal articles, and on walks along the shores of Lake Mendota in Madison, Wisconsin, where I lived.

When I returned from Bolivia a year later, June asked me to serve as co-editor, a role that I would soon learn required the skills of negotiator, businessperson, copyeditor, and judge. In the beginning of my efforts, I undoubtedly embodied the truth of Ambrose Bierce's description of the editor as "a gilded impostor." Nevertheless, necessity forced me to quickly learn the art (not science!) of editing a volume of essays, as well as the steps of the delicate dance that editors must perform with publishers.

Even with a steep learning curve, I could not have moved an inch forward without the constant guidance of June and the volume's contributors, whose collective experience and wisdom in matters subtle and (to them, at least) not-too-subtle, gave me a fighting chance. Despite her leading role, by early 2001 June's attention had turned to more profound matters. My hope is that June would have been satisfied with the form the volume finally assumed. Her intellectual vision for an enduring ethnographic study of law will continue to remain relevant and vital.

I would like to thank our editors at Palgrave Macmillan, particularly Kristi Long and Sonia Wilson, whose enthusiasm and expertise smoothed the process.

My research assistant at Emory University, Leslie Dyer, provided timely and accurate support during production.

Finally, I must acknowledge the sustaining presence through all of this of Romana and Dara—my muses.

M. Goodale

PREFACE

June Starr

June O. Starr, who passed away on April 27, 2001, is the intellectual author of this collection. As an able and dedicated ethnographer, she became concerned in the mid-1990s that many of the social and legal scholars who were advocating "ethnography" as a route to knowledge did not have the tools and training to realize their goals. She thus wrote to several of her friends and colleagues, including us, to suggest that we who had done ethnography write about our experiences and methods. She asked us to contribute papers for two panels, to be held at the 1998 meeting of the Law and Society Association in Aspen, Colorado. The panels were well attended, and some of the papers presented there have been revised for inclusion in this volume.

The project expanded after the Aspen meeting, as good projects do. Many of us who had contributed papers joined June in realizing that a volume on ethnographic methods for studying law would be useful, not only for sociolegal scholars in other disciplines but also for anthropologists, such as ourselves, who were training students in the field. We had all shared stories of the clever methods we had designed to answer questions that arose in our fieldwork. But nowhere were these stories collected into a volume that we could give to students and colleagues planning research. June helped us to see that our methods, although sometimes designed for unique situations, could nevertheless prove useful to others. Future scholars could both mine our essays for methods they might use in their own research and learn the important lesson that doing ethnography, in addition to creativity, requires hard work, time, and attention to detail.

Sadly, June's health declined as the project grew. In the summer of 1999, she enlisted the help of Mark Goodale, who worked with June in collecting and editing papers and writing an introduction. After she passed away, he took over the project and brought it to the point of publication. Through Mark's efforts, June's project has finally come to fruition. This volume, which is dedicated to her, reflects June's vision that good scholarship requires good methods, however various these may be. It reflects her concern that ethnography in law and society research be intensive and long-term, and her conviction that there can be no substitute for careful, empirical research.

June Starr's own career as an anthropologist and legal scholar reflects her long-standing emphasis on the role of ethnography in sociolegal studies. Her legacy in the field of legal anthropology is an important and wide ranging one. She has contributed in significant ways to the development of the field as a whole. Her early work began with the study of dispute processing inspired by Laura Nader's emphasis on the social processes of conflict rather than the content of rules. She received her Ph.D. at the University of California, Berkeley, where she was a member of the Berkeley Village Law Project, directed by Laura Nader. She moved from this work to the analysis of history and power in the intersections of multiple legal systems and to the dynamics of legal change over time. Most of her work focused on Turkey, although she also studied mediation in the United States and legal relations in India. She was on the faculty at SUNY-Stony Brook from 1970–1995, then at the Indiana University Law School in Indianapolis until her retirement in 2000.

Her first book, *Dispute and Settlement in Rural Turkey: An Ethnography of Law* (1978, Leiden, Netherlands: E. J. Brill), was a masterful study of disputing in a Turkish village, inspired by Laura Nader's effort to reshape the field of legal anthropology. Disputing research sought to turn legal anthropology from sterile discussions about the nature of law to a focus on legal behavior in all its manifestations. Disputing was the central unit of analysis, and June provided a wonderful study of how the residents of a Turkish village handled conflicts. Turkey was her first and abiding intellectual love, even though she worked in other areas of the world. As a young scholar, she quickly became caught up in the excitement of the early 1980s about a new approach to conflict resolution called "alternative dispute resolution," or ADR. ADR was of particular interest to anthropologists because it was an attempt to transplant forms of conflict management from the kinds of village societies around the world that anthropologists such as June had studied in detail, to urban neighborhoods in the United States diagnosed as overly litigious and lacking a sense of community.

But June saw the limitations of a focus on dispute processing. Once the shape of disputes was analyzed, there was the need for a more structural analysis: one that paid attention to the law itself. She started to read about Sir Henry Maine's theories of law, about the law merchant in the Ottoman Empire, and about the history of the formation and development of legal systems. At this point she became involved with the Commission on Folk Law and Legal Pluralism and interested in analyzing situations in which more than one legal system coexisted, often as a product of colonialism. She participated in the 1988 meeting of this group in Zagreb, part of the International Union of Anthropological and Ethnological Sciences, and expanded her work on the development of legal systems.

June's concern with legal systems rather than disputes, and her interest in examining the dynamics of legal pluralism over time within shifting power relations, inspired her to organize a conference on the state of legal anthropology. The conference was held in Bellagio, Italy in 1985 and the papers were later published in an edited volume, *History and Power in the Study of Law: New Directions in Legal Anthropology* (1989, Cornell University Press, June Starr and Jane Collier, eds.). This important and widely read collection broke new ground in its move from the study of disputing as a social process to the examination of law as an institution and as a cultural system. The book called for more attention to power relations and to historical perspectives on law in society. It was, in a sense, a critique of the disputing approach. The book did not advocate abandoning the analysis of disputing, but suggested the need to consider power and history as well. At the same time, it criticized legal pluralism for an overly static analysis of the intersection of legal systems. Indeed, June's next book was historically focused. Her interest in understanding power and legal change resulted in her elegant analysis of changing law in Turkey,

Law as Metaphor: From Islamic Courts to the Palace of Justice (1992, SUNY University Press). She was always interested in gender relations as well, and published several important articles on this topic, including two on gender and social change in the *Law and Society Review* (1974; 1989).

June remained committed to ethnography, concerned that the field of anthropology was becoming too postmodern, too distant from its roots of trying to understand the culture and social life of other societies through careful empirical research. She thus conceived of this edited volume in the mid-1990s as a way of articulating the methods used in legal anthropology and making them more widely known to other sociolegal scholars. She sought out friends and colleagues whose ethnographic work exemplified the commitment to empirical research that she advocated and invited them to contribute papers for the panels she was organizing for the Law and Society meetings in Aspen. At those meetings she emphasized the extent to which theory must be rooted in the practice of ethnography.

June followed her love of law by attending law school and then by teaching in law school, after a long career of teaching anthropology. She loved her students and found teaching them a gratifying experience. Her law school teaching career was cut short by her deteriorating health, but it is typical of her that she was full of new ideas for things to study, from biodiversity and its legal regulation to the creation of trusts for elderly people. She found the world full of fascinating issues. But she was also wary of jumping on the next bandwagon, of abandoning her roots in ethnography. She was a leader in many ways, both in her intellectual contributions to the anthropology of law and in her own life, following her passions for learning about law and teaching others. She was an important member of the Law and Society Association, closely connected with it for twenty-five years, and an important leader in the field of legal anthropology. She helped to turn legal anthropology toward a more institutional understanding of law while retaining its focus on careful ethnographic research. This book reflects her vision.

Jane F. Collier
Sally Engle Merry

INTRODUCTION

LEGAL ETHNOGRAPHY: NEW DIALOGUES, ENDURING METHODS

June Starr and Mark Goodale

This volume is the result of a project in which ethnographers were asked to write essays that explored different research techniques that they have used to understand complex legal issues in a variety of social, political, and legal contexts. Each essay addresses a particular set of analytical problems that the ethnographers have confronted, or a particular type of data or research method that the ethnographer found useful. Because of its detailed focus on actual research techniques by ethnographers studying law in diverse settings, this volume will serve as a guide for students who are designing their own methodologies, for scholars who are new to, but interested in the possibilities of, ethnographic research, and for experienced ethnographers who are interested in theorizing about or reflecting on their own particular methodological problems. Like all ethnographic research, ethnographic studies of legal phenomena require both a serious engagement with theory and a methodological rigor. The chapters in this volume demonstrate how individual scholars have faced this two-part challenge in the course of research in different parts of the world and during different time periods.

Theory and methods are mutually constitutive, but the process by which each constitutes the other is not a simple or static one. As the chapters in this volume show, legal ethnographers embark on research projects after carefully considering prevailing theory and after reflecting on the research methods that would best accomplish research goals in light of theory. But part of what makes the interaction between theory and methods so dynamic is that researchers inevitably encounter new challenges during the research process itself, and these new challenges lead researchers to rethink both theoretical assumptions and methodological practices. The trajectory through which theory and methods develop is not unilineal; in each generation, scholars find that certain ethnographic methods—like participant observation and interviewing—remain useful, and indeed necessary, while the concrete problems of the research process itself force researchers to experiment and adapt their research plans in new and unpredictable ways. After completing research projects and during the process of reflection and writing, scholars debate which theory and methods continue to prove useful and what new variations on theory and methods should be seriously considered. Readers of the chapters in this volume will be able to vicariously participate in this process as scholars from different disciplines, who use ethnographic methods to study legal phenomena, engage in these ongoing debates.

The Value of Legal Ethnography

Many legal contexts today continue to be dynamic or unstable in unpredictable ways. For example, the decision-making processes of many institutions, especially international tribunals, are subject to nonlegal political influences from transnational sources such that legal institutional integrity is commonly called into question. Ethnographic methods are useful tools for accessing the complex ways in which law, decision-making, and legal regulations are embedded in wider social processes. Ethnographic methods are also useful to scholars who employ a historical framework or who examine legal change. As both Sally Engle Merry and Lawrence Friedman describe in their chapters, ethnographic methods can be used to build a contemporary context in which to understand legal history.

Scholars who study the sociological context of law continue to develop traditional ethnographic approaches and their projects continue to integrate the historical study of an institution or a group into the fundamental fabric of research projects. Scholars also continue to engage in dialogues across disciplines about the effectiveness of different methods, in light of concrete research problems. Moreover, the degree to which a researcher employs qualitative or quantitative techniques is dictated in large part by the exigencies of the research project itself, the context of the research, and new questions that emerge in the course of research.

Researchers who are new to ethnography should approach this endeavor with rigor and with an understanding of the types of commitment it demands. Ethnographic research requires a serious grappling with complex methodological and theoretical issues. Although anthropologists have played a fundamental role in the development of ethnographic methods, other social sciences have, especially in recent years, begun to see the usefulness of ethnographic techniques. Researchers interested in taking an ethnographic approach to the study of legal phenomena should keep in mind the fact that ethnographic research is more than simply "hanging out" with legal actors; rather, the effective use of ethnographic techniques, like participant observation and interviewing, requires in most cases prior training and careful planning (see Nader in this volume). Even though it is not discussed very often today, we should also underscore the fact that good ethnographic research requires a significant time commitment. Despite the fact that the actual length of time will vary depending on the project, most ethnographers working with different methodologies and theoretical orientations have found that a minimum of twelve months is necessary to capture the kind of rounded picture needed for later analysis.

Moreover, as several of the chapters in this volume show—especially those by Collier, Coutin, and Hirsch—there are, on top of the practical difficulties associated with the long-term use of ethnographic methods, significant ethical issues raised by the use of these methods within highly charged legal settings. Legal ethnographers are often, in the course of research, given access to legal knowledge of great consequence for people, and the use of ethnographic methods demands that the researcher carefully consider the impact of the production of knowledge on research participants. Ethnographic research tends to produce knowledge that is deeply intertwined

with local power structures. This means that legal ethnographers often find themselves confronting difficult problems, the political, legal, and methodological aspects
of which are often impossible to untangle.

The Project of Ethnographic Research

A central theme that emerges from the articles in this volume is the notion that
ethnography is a project that develops over time. Legal researchers in the field
encounter contexts that could not be predicted in advance and that challenge fundamental methodological assumptions. This uncertainty is to be expected and means
that ethnographers should remain flexible and adaptive. But the project of ethnography does not end when research does; rather, the researcher also encounters new
theoretical literatures that challenge working assumptions about the nature and
meaning of research findings.

For example, recent critical approaches to the tensions and conflict generated by
the globalization of communication networks, transportation systems, and trade
(see, for example, Appadurai 1996), suggest that strategies from the ethnographic
tradition can document the ways people react to or even resist the march of global
integration (see, for example, Eve Darian-Smith's work on the "chunnel" between
England and France; see also Bryant Garth and Yves Dezalay's examination of the
how the "transnational legal order" impacts various South American countries). In
conceptualizing globalization, Appadurai places more emphasis on the ways people
around the world resist globalization through their imagination. Legal ethnographers
also analyze resistance in the form of local social movements, acts of civil disobedi-
ence, cases in local courts, and cases before international legal forums. These are
arenas that demand the special training of ethnographic researchers.[1]

A key dilemma for legal ethnographers is how to study local level phenomena that
become "global-legal" events. Although this is not the only solution, some legal
ethnographers have begun to adopt a "multisited" focus in order to view the various
points and opinions which make up complex networks (see, for example, Goodale
in this volume; see also Merry 2001).[2] As Marcus (1995, 1998) and others have
argued, "multisited research"—in which an ethnographer does research in several
sites—is one way to comprehend the linkages that form the "world system" and
constitute what was formerly understood as "the local system."

This conceptualization of legal ethnographic work in broader spatial terms will
continue because it allows a researcher working to achieve finely grained ethnographic detail to gain insight into the importance of regional and global legal
processes. Thus, studies of "intersecting convergences" emerge, such as how the "car
and train tunnel" under the channel between England and France is affecting the
cultures on each side (Darian-Smith 1999), or how the multicultural genotype
projects are affecting scientific research (Rabinow 1996, 1999).

Different Disciplines, Common Themes: A Survey of Methodologies

Because this volume brings together legal ethnographers from different disciplines—
anthropology, sociology, political science, and law—it is not surprising that different
contributors focus on different parts of their ethnographic research. What emerges is

an impressive array of research methods. These methods are complementary sticks in an expansive bundle of methodological options. Legal ethnographic research, like all ethnographic research, is enriched by this diversity. Moreover, the kind of interdisciplinary dialogue regarding methods on display here, centered around common themes, is an important part of the project of ethnography.

In this last section of the Introduction, we introduce the chapters in two ways: First, we give a brief synopsis of each so that the reader can glimpse each chapter's content; second, we arrange the chapters thematically, highlighting important differences and similarities.

The Chapters

The volume is divided into two parts. Part I is entitled "Performing Legal Ethnography" and the bulk of the chapters are found here, which is the heart of the volume. Each chapter focuses on the specific techniques used by the researcher in the course of designing and conducting a legal ethnographic research project in light of the theoretical approaches the researcher found relevant and useful.

Susan Hirsch's chapter is based on fieldwork in Tanzania that involved her with local activist groups organizing for broader legal rights for women. Hirsch argues that legal ethnographers can learn from feminist researchers who regularly blur the line between research and personal engagement. Specifically, she contends that borrowing methods from wider feminist scholarship and activism can help legal ethnographers negotiate "the methodological difficulties of studying legal consciousness; the problematic relationship between the researcher and research subject in ethnographic fieldwork; and the blurred boundaries between scholarship and activism, especially in sociolegal studies."

Philip Parnell makes innovative use of media studies in the course of his research. He argues that the legal ethnographer should try to discover another culture's nonacademic theoretical inventions by conducting research into that culture's thoughts and actions. A study of disputing is, according to him, a very good way to do this. To place his experiences in an urban squatter settlement in Manila in perspective, he examines the ways that law is communicated through the media in the United States. Parnell suggests that the way an ethnographer understands legal roles in the United States will affect the researcher's perceptions of another legal culture. Parnell found that in order to understand urban legal culture in Manila, he was required to "trek" up and down the informal legal networks that connect the state with people at the bottom of the national legal hierarchy.

Mark Goodale's article discusses the implications for conducting legal ethnographic fieldwork in an era of globalization. Based on fieldwork conducted in the late 1990s, this article describes several ways in which legal practices and ideas have become globalized in new ways. The article uses several aspects of fieldwork in Bolivia to demonstrate the ways in which a legal ethnographer can respond to these global contexts by adopting, among other methods, a multisited approach to research. In the late 1990s a series of events in Bolivia resulted in the penetration of Western human rights discourses into many rural areas. In one striking example of this process, legal services centers were opened in rural Bolivia that were supposed to protect the rights of women and children, as outlined in specific United Nations and related proclamations. The directors of these centers traveled to various cities in Bolivia to receive training in these rights doctrines from both urban intellectuals and representatives of international agencies, after which they returned to their villages to educate locals. The introduction of these new understandings of human rights led to both ideological and practical transformations at the local level. The author was able to track these global, national, regional, and local articulations by re-envisioning the research site as multileveled and multidimensional.

Jane Collier shows the usefulness of sociolinguistic analysis as a method in the course of her research on witchcraft beliefs in Chiapas, Mexico. Collier argues that in her experiences as a legal ethnographer, traditional methods that led her to ask questions about crimes and punishments were inadequate to explain why people divorced or sought redress in the local court for a variety of harms. This discovery was facilitated in part by analyzing the specific terms local people used in their own language for the various offenses. This analysis revealed that legality, witchcraft, and religion are not separated in people's minds in Chiapas. However, although religion and law are interrelated, Collier rejects the notion that "law," broadly defined, serves primarily as a means of social control, a view that, she suggests, reflects Western biases. Her use of sociolinguistic analysis implies that people in Chiapas are engaged in a constant process of negotiation, a process that does not finally resolve conflicts, but only manages them from generation to generation.

Robert Kidder's chapter's focuses on the usefulness of the comparative method, and is based on field research in Japan and among the Amish in Lancaster County, Pennsylvania. The received wisdom regarding people of both cultures is that they value "law-avoidance," meaning that they employ various means to avoid having to resort to legal arenas to resolve conflicts. Kidder's attempt to test this received wisdom by exploring the legal consciousness of both cultures raises three major methodological issues. First, how should the researcher interpret statements by people who reject comparative analytic approaches to conflict? Second, how can the statements and actions of both Japanese and Amish be interpreted reliably and validly? And, third, how can ethnographic research be both a vehicle for the subjective "voice" of the research subjects, and yet contribute to generalized social scientific knowledge?

Susan Coutin's chapter explores methods used during her ongoing fieldwork at the borders of immigration and politics in California. Because her work involves her at the center of intense political and legal issues in California, she demonstrates the fact that participant observation and interviewing can never be simply neutral techniques for gathering data. Rather, such activities by the researcher place her "among a group of people who are engaged in a particular set of activities." In Coutin's case, these activities are related to political and legal activism in support of immigrants. An important theoretical implication derived from Coutin's work is that particular research contexts can give new (and sometimes politicized) meanings to traditional ethnographic techniques.

Sally Engle Merry explores the implications of "doing ethnography" in legal archives in Hawai'i. There are several challenges confronting sociolegal researchers attempting to use legal archival material. As Merry shows, an anthropologist using archival legal texts, such as court records, faces a challenging situation: In order to make sense of these records, they need to be grounded in an ethnography of the surrounding community. This includes an analysis of its major actors and its changing political, economic, social, and cultural terrain over a long period of time. In order to do this, the anthropologist must take an ethnographic approach in the archives. As Merry argues, ethnographic work and archival research naturally compliment each other, and both should be used by the sociolegal researcher, especially when one is attempting to understand legal processes and social change over time. Merry is skeptical that one technique can be used without the other when addressing questions of long-term social and legal change. Legal ethnography without archival research cannot make connections between court processes and the changing social order; conversely, archival research without social history risks capitalizing only on dramatic trials at the expense of the everyday social order and it may ignore the local contextual relationships which an ethnographic perspective can reveal.

Herbert Kritzer's article is a critical discussion of the limitations of interviewing as an ethnographic method compared with participant observation. His research in Madison, Wisconsin consisted of the following: a structured mail survey of Wisconsin attorneys (working on a contingency fee basis), observations in three different firms (one month in each firm), and a series of semi-structured interviews with contingency fee practitioners around the state. Using extended portions of observational field notes in his chapter, Kritzer concludes that observation allows the researcher to gather a qualitatively different kind of data compared with interviewing. Because of the open-endedness of observation, the ethnographer is able to discover patterns of behavior, in this case patterns among practicing attorneys, that the interview setting simply cannot uncover. For Kritzer, the optimal research project would use a combination of both interviewing and observation, with observation serving as the "check" against the data gathered from initial and ongoing interviews.

Anne Griffiths's research involved the extensive use of life histories as part of a wider ethnographic study in Botswana on women's difficulties in using the legal system, or in trying to resist it. Only by analyzing life histories did she understand the extent to which legality in Botswana is an essentially gendered process, and reflects the gendered nature of society as a whole. Within the legal system, Griffiths found there were even differences *between* women in terms of the amount of protection they could expect from law: These differences reflected power differentials between women, based on factors that are not usually acknowledged in feminist debates on law. Besides illustrating the effective use of life histories in legal ethnography, Griffiths's paper is an important contribution to feminist legal debates.

Part II of *Practicing Ethnography in Law,* "Reflections on Ethnography in Law," rounds out the volume with two contributions that reflect on the past, present, and future of legal ethnographic work. The first article, by Lawrence Friedman, a law professor and legal historian, muses on the value of a cultural anthropological approach, and in doing so he responds to some recent criticisms of the field. In particular, he discusses the ways that the analysis of court records, collected over time, need to be put in a cultural context to be understood. Friedman has conducted legal archival work on a wide range of topics, from how court reform has attempted to protect senior citizens to American inheritance patterns. Here he draws largely from his work on changes in American divorce law from the late nineteenth century to the present. His essay is an affirmation not only of the value of ethnographic methods for legal history, but the *necessity* for them. He argues that raw statistical data are meaningless without placement in a wider social context.

The volume concludes with a chapter by Laura Nader. Nader's contribution, written in the style of what she calls "an intellectual autobiography," has one major purpose: to show that legal ethnography, as developed and practiced by legal anthropologists this century, is not synonymous with fieldwork, but rather describes flexible and changing approaches to the study of sociolegal research problems. As Nader argues, although legal anthropologists have found participant observation, interviewing, and other classic techniques indispensable in pursuing ethnographic research, the use of such techniques does not make a fieldwork project an ethnography *as such*. She uses examples from her many different research projects—beginning in 1957—to illustrate the her notion that "in ethnography, the methods are subordinate to the questions being pursued. Methods become eclectic because a loyalty to a single technique, even something like participant observation, commonly stultifies research."

Common Themes

Besides the fact that all of the chapters in the volume demonstrate the use of diverse methods as part of the project of ethnography, there are several specific convergences between the various chapters in *Practicing Ethnography in Law*. The chapters by Coutin, Griffiths, and Hirsch provide examples for how researchers can design methods which allow them to take an active role in the institutions and phenomena they study. This active personal engagement will often be necessary for ethnographers studying legalities linked to wider social issues of power, gender, and marginality; and the contributions by Coutin, Griffiths, and Hirsch show some ways in which research methods can be adopted to facilitate this important process.

The chapters by Goodale, Kidder, and Parnell show how methodologies must now take into account the local–regional–global articulations that are amplified by processes of globalization. Kidder and Parnell both confront this problem through the use of the comparative method. In doing so, both show the difficulties with arriving at theoretical generalizations while attempting to stay true to the subjective voice that is so important in ethnographic research. Goodale, on the other hand, describes a possible solution to researching globalized legal phenomena that depends on a reformulation of the meaning of the "fieldsite," as well as a willingness, like Parnell, to "trek" up and down legal structures wherever they lead.

Several of the chapters explore the advantages and disadvantages of quantitative and qualitative methods. The chapters by Friedman, Kritzer, and Merry suggest that legal ethnographic projects should use both quantitative and qualitative methods when possible. Merry and Friedman focus on the advantages to be gained from investigating legal archives and from embedding the quantitative findings in the ethnographic contexts from which they were derived. Kritzer focuses on the benefits and disadvantages of interviewing people versus observation of how they behave, two classically qualitative methods. He then locates these methods in relation to the more quantitative techniques that he is familiar with as a political scientist.

All the ethnographers in the volume have critically reevaluated specific methods they have used in the course of their research projects. This is, indeed, an important part of the project of ethnography itself. Collier's article, for example, shows how she came to realize that her earlier focus on legal processes did not help her understand why local people used the courts. By changing her approach to examine the language people used while in court, she gained deeper insight into law's meaning in Chiapas. Finally, Nader draws on her research experiences since the late 1950s to show how she has continually critiqued and reformulated her own research techniques, even as the theoretical problems she confronted have changed.

Conclusion: Legal Ethnography and Its Possibilities

What the chapters in *Practicing Ethnography in Law* demonstrate above all is that even as the theory behind legal ethnographic research shifts as part of the ongoing project of ethnography, some research techniques will remain stable and effective. This continuity allows researchers to view the whole sweep of sociolegal scholarship with critical and constructive eyes in order to develop an appreciation for the enduring problems of law and society research and to understand that deep and thick ethnography is one of the best routes we have in comprehending the complexity of law and legal processes in a changing society.

Notes

1. For other examples of the ways in which legal ethnographers have grappled with the problems of globalization during research, see Goodale 2001; Maurer 1997; Merry 2000; Riles 2000; and Woodiwiss 1998. For a critical discussion of the issues facing legal ethnographers working in a globalized context, see Darian-Smith 2000; Nader 1995, 1999. For a discussion of the challenges facing researchers working in contexts with NGOs and other development agencies—but symbols of globalization—see the recent symposium on "NGOs, Power, and Development" in the *Political and Legal Anthropology Review* (PoLAR 2001).
2. Other ways to research global-legal events that can unfold simultaneously in dispersed locations include the use of collaborate research teams and the use of information technology, like the Internet, which can be used to communicate with informants and conduct research, at the same time.

References

Appadurai, Arjun
 1996 Modernity at Large: Cultural Dimensions of Globalization. Minneapolis: University of Minnesota Press.
Darian-Smith, Eve
 1999 Bridging Divides: The Channel Tunnel and English Legal Identity in the New Europe. Berkeley: University of California Press.
 2000 Review Essay: Structural Inequalities in the Global-Legal System. Law and Society Review. Vol. 34.
Dezalay, Yves and Bryant Garth
 1996 Dealing in Virtue: International Commercial Arbitration and the Construction of a Transnational Legal Order. Chicago: University of Chicago Press.

Dezalay, Yves and Bryan Garth
    1998   Chile, Law and the Legitimation of Transitions: From the Patrimonial State to the International Neo-liberal State. Chicago: American Bar Foundation. Working Paper No. 9709.
French, Rebecca
    1995   The Golden Yoke: The Legal Cosmology of Buddhist Tibet. Ithaca: Cornell University Press.
Goodale, Mark
    2001   A Complex Legal Universe in Motion: Rights, Obligations, and Rural-Legal Intellectuality in the Bolivian Andes. Doctoral dissertation, Department of Anthropology, University of Wisconsin.
Harrington, Christine and Barbara Yngvesson
    1990   Interpretive Sociolegal Research. Law and Social Inquiry 15(1).
Lazarus-Black, Mindie and Susan Hirsch
    1994   Contested States: Law, Hegemony and Resistance. New York: Routledge.
Marcus, George
    1995   Ethnography in/of the World System: The Emergence of Multisited Ethnography. Annual Review of Anthropology 24: 95–117.
    1998   Ethnography through Thick and Thin. Princeton: Princeton University Press.
Maurer, William
    1997   Recharting the Caribbean: Land, Law, and Citizenship in the British Virgin Islands. Ann Arbor: University of Michigan Press.
Merry, Sally Engle
    1990   Getting Justice and Getting Even: Legal Consciousness Among Working Class Americans. Chicago: University of Chicago Press.
    2000   Colonizing Hawai'i: The Cultural Power of Law. Princeton: Princeton University Press.
    2001   Crossing Boundaries: Methodological Challenges for Ethnography in the Twenty-first Century. Political and Legal Anthropology Review (23)2.
Moore, Sally Falk
    1973   Law and Social Change: The Semi-Autonomous Field as an Appropriate Subject of Study. Law and Society Review 7: 719.
    1978   Law as Process: An Anthropological Approach. London: Routledge and Keegan Paul.
Nader, Laura
    1990   Harmony Ideology: Justice and Control in a Zapotec Mountain Village. Stanford: Stanford University Press.
    1995   Civilization and its Negotiations. In P. Caplan (ed.) Understanding Disputes: The Politics of Argument. Providence, R.I.: Berg.
    1996   Naked Science: Anthropological inquiry into Boundaries, Power, and Knowledge. New York: Routledge.
    1999   The Globalization of Law: ADR as "Soft" Technology. Proceedings of the 93rd Annual Meeting of the American Society of International Law. March 24–27, 1999, Washington, DC.
PoLAR
    2001   Symposium, "NGOs, Power, and Development." Vol. 24, No. 1.
Rabinow, Paul
    1996   Making PCR: A Story of Biotechnology. Chicago: University of Chicago Press.
    1999   French DNA: Trouble in Purgatory. Chicago: University of Chicago Press.
Riles, Annelise
    2000   The Network Inside Out. Ann Arbor : University of Michigan Press.

Santos, Boaventura de S
1987 Law: A Map of Misreading; Toward a Postmodern Conception of Law. Journal of Law and Society 14: 279.
1995 Toward a New Common Sense: Law, Science and Politics in the Paradigmatic Transition. New York: Routledge.
Sarat, Austin
1990 Off to Meet the Wizard: Beyond Validity and Reliability in the Search for a Post-empiricist Sociology of Law. Law and Social Inquiry 15(1).
Starr, June
1978 Dispute and Settlement in Rural Turkey: An Ethnography of Law. Leiden: Brill.
1989 The Role of Turkish Secular Law in Changing the Lives of Rural Muslim Women, 1950-1970. Law and Society Review 23(3): 497–523.
1992 Law as Metaphor: From Islamic Court to the Palace of Justice. Albany: State University of New York Press.
Starr, June and Jane F. Collier
1989a History and Power in the Study of the Law: New Directions in Legal Anthropology. Ithaca: Cornell University Press.
1989b Dialogues in Legal Anthropology. In History and Power in the Study of the Law: New Directions in Legal Anthropology. J. Starr and J. Collier, eds. Ithaca: Cornell University Press.
Starr, June and Jonathan Pool
1974 The Impact of a Legal Revolution in Rural Turkey. Law and Society Review 8(4): 533–560.
Stocking, George
1991 Observers Observed: Essays on Ethnographic Fieldwork. Madison: University of Wisconsin Press.
Thompson, E. P.
1990 [1975] Whigs and Hunters: The Origin of the Black Act. London: Penguin Books.
Trubek, David
1990 Back to the Future: The Short, Happy Life of the Law and Society Movement. Florida State University Law Review 18(1): 55.
Woodiwiss, Anthony
1998 Globalisation, Human Rights, and Labour Law in Pacific Asia. Cambridge: Cambridge University Press.

PART I
PERFORMING LEGAL ETHNOGRAPHY

Chapter 1

Feminist Participatory Research on Legal Consciousness

Susan F. Hirsch

Introduction

Many studies in legal anthropology examine gender relations, yet virtually absent from the literature is an engagement with the debates over feminist method that have shaped the study of gender in other disciplines (see, e.g., Harding 1987; Reinharz 1992) and in other anthropological subfields (see, e.g., Bell, Caplan, and Karim 1993; Jackson 1986; Strathern 1987). In writing about feminist method as part of a broader consideration of methods in legal anthropology, my concern is neither to rehearse debates over feminist methodology and epistemology outside the subdiscipline, nor to offer a blueprint for deploying feminist methods in legal anthropology, nor even to advocate that scholars studying gender and law should take a feminist approach to their research. Rather, by describing a research project that incorporates feminist methods—namely, a workshop on legal consciousness conducted with a feminist activist group in Tanzania—this chapter illustrates how a consideration of feminist methods might help legal anthropologists to negotiate, or at least to reflect on, several thorny issues in contemporary research: the methodological difficulties of studying legal consciousness; the problematic relationship between the researcher and research subject in ethnographic fieldwork; and the blurred boundaries between scholarship and activism, especially in sociolegal studies.

The feminist participatory workshop analyzed in this chapter was conducted as part of a research project on gender and legal change in Tanzania. My description of that project's development in the next two sections situates it in relation to recent anthropological and feminist debates over research strategies and the politics of ethnography. Subsequent sections analyze the workshop itself, drawing out the specific lessons learned with respect to legal consciousness and to subjectivity in the research process. The concluding sections also address the relation between studying legal consciousness and attempting to change it, as confronted by the scholar/activist engaged in research.

Developing a Feminist/Activist Research Project

The research project discussed in this chapter was designed in part to overcome limitations of earlier ethnographic research on law and gender that I conducted in coastal

Kenya (see Hirsch 1994, 1998). Although satisfied with my conclusions concerning law's role in shaping gender relations, I was dissatisfied with aspects of the research experience, especially the disjuncture between my goals as a researcher and those of the people whose lives were my focus. My main goal was to analyze cases, mediations, and conversations about matrimonial problems to demonstrate that Muslim law, like other forms of law, was used actively by laypeople and legal personnel both to maintain and to transform gender relations. The example highlights the diversity of Muslim practice and thus challenges stereotypes of Muslim societies. Many of the Kenyan Muslims who participated in my research recognized the relevance of this goal in a largely anti-Islamic world; however, they were at times frustrated that their own goals of representing the community in positive, even idealized, ways were less important to me. My very focus on marital problems and court cases risked exposing shameful family secrets that many people would have rather kept hidden (Hirsch 1998). A second dissatisfaction with the Kenyan project was my tendency to mask or subordinate my political views as a feminist and also as someone opposed to the Kenyan government's oppression of dissidents, which was rampant at the time. Though rarely expressed to Kenyans, those views implicitly informed my conduct as a researcher (Coutin and Hirsch 1998). Thus, in developing new research on gender and law in Tanzania, I resolved to represent myself more openly as a political academic and to incorporate more effectively the goals of other participants in the research process.

My desire to design a very different kind of project stemmed not only from the field experiences discussed above but also from reading an emerging literature on feminist methods (see, e.g., Harding 1987; Gluck and Patai 1991). Shulamit Reinharz identifies themes that characterize the wide range of feminist research. According to Reinharz, "guided by feminist theory" and "transdisciplinary" feminist research "aims to create social change" and "strives to represent human diversity" (1992: 240). The characteristics of feminist research that it "frequently includes the researcher as a person" and "frequently attempts to develop special relations with the people studied (in interactive research)" were especially relevant to my research plans. Feminist research realizes these themes through revising mainstream methods and developing new methods, including oral history, personal narrative, and interactive forums for collecting data. In some projects, the researcher's authority to determine project design, data collection techniques, and other aspects of research is shared with those participating as research subjects. Thus, techniques of feminist research routinely question and blur the boundaries between observer and observed. Moreover, the goals of all participants, including their political goals, are of critical concern in selecting methods.

A related literature emerging in the early 1990s engaged similar issues by questioning the neocolonial aspect of much feminist anthropology (Mohanty 1991; Mihn-Ha 1989). The call for Western feminist social scientists to examine their privileged positions in the construction of knowledge about women and gender cross-culturally, articulated some of my discomfort with the Kenyan project. Specifically, these writings exposed and seriously questioned the power dynamics that inhered in First World women conducting research on Third World women, arguing that much feminist writing reified these categories and reinforced the power relations reflected in them. Although both categories, and the relation between them, are more often stereotyped than adequately theorized, even in critiques, my own position in the first of them compelled me to confront the charges directly through reasoned and politicized choices of research topic and method.

In response to the scholarship and experiences described above, I developed a project intended to be relevant to all participants and conscious of the power dynamics inherent in our relationships to one another. The project, which examined legal consciousness and law reform examines the growing concern—in Tanzania and globally—over fostering consciousness about women's legal rights. This choice of topic allowed me to conduct research with people acting on political goals with respect to gender relations, such as activist lawyers and members of women's organizations. Studying their efforts to raise legal consciousness and to effect change required me to develop new research methods. Focusing on legal consciousness also forced me to grapple with my positioning as a researcher. My reading of Mohanty's trenchant critique of the cross-cultural study of Third World women by Euro-Americans suggested that researchers try to find links or common projects with women in Third World contexts rather than recreating us/them distinctions. Moreover, given the climate of intensive law reform, I would need to confront explicitly my political position as a foreign researcher. What would my role be as those around me—including many non-Tanzanians—engaged in raising legal consciousness? I welcomed these challenges but, to be frank, found few examples in legal anthropology of specific methods for politically engaged research. My approach therefore is to forge a reconstruction of activist anthropology by studying the process of how consciousness is shaped by and in relation to discourses and practices in which I am an acknowledged participant.

Studying Legal Consciousness and Law Reform

Legal consciousness is constituted in relation to the legal processes available to people, the ideas and practices of legal professionals and laypeople, and also discourses circulating locally and internationally (see, e.g., Merry 1990; Mraz 1997).[1] When I began research in Tanzania in 1992, efforts to reform law included attention to women's rights and also broader demands for human and civil rights. Throughout the 1990s, my research has tracked a tremendous increase in the production of such discourses by governmental and nongovernmental bodies (e.g., women's organizations), the good governance programs of donor nations, and projects undertaken by legal personnel, scholars, activists, and individuals. All of these efforts operate on local, national, and international levels that intersect in interesting ways. In terms of theory, the research explored the impact of certain global discourses in local contexts, focusing on how institutions of nation-states (e.g., the courts) and civil society (e.g., women's organizations and religious bodies) make use of internationally circulating ideas and practices and, at the same time, transform them to address local prerogatives (cf. Wilson 1995). This global/local interaction is crucial to the development of legal consciousness among people at all levels of society. The challenge is to identify which local imperatives, structures, and processes shape how these discourses meet in a given context and thereby how they influence consciousness.

Initially I conducted research using a variety of standard methods for the ethnographic study of legal consciousness. For example, I observed cases and reviewed hundreds of case files in a Primary Court in a small town and also in Magistrates' Courts in the large city of Dar es Salaam.[2] I combined this review, which revealed how people used the law, with interviews of legal personnel, law reform advocates, and laypeople with court experience. The specific understandings that people have of law are a central aspect of legal consciousness, and such understandings, as revealed through interviews and casual conversations, are profitably analyzed in relation to patterns of court use. In my research I also consulted relevant scholarly and legal documents. I interpreted the discourse used in legal cases and documents as well as in conversations about law. These methods of investigating legal consciousness were invaluable to the project; however, they did not address my concern to pursue politically engaged research. Accordingly, I directed attention more specifically to those individuals and groups who were working to change legal consciousness.

The tactics and discourses of several organizations engaged in campaigns for law reform and legal consciousness became a focus of my research.[3] Through educational campaigns, TAMWA (Tanzania Media Women's Association), SUWATA (a woman's organization affiliated with the ruling party), the Legal Aid clinic of the University of Dar es Salaam Law Faculty, TAWLA (Tanzania Women Lawyer's Association), TGNP (Tanzania Gender Networking Programme) and other organizations have been seeking to create "legal literacy"—their phrase—in domestic violence, rape, sexual harassment, inheritance, widow inheritance, and the regulation of women's labor. Their central concern has been to acquaint the public, especially women, with legal rights, and their efforts have mushroomed with donor funding in the late 1990s (Hirsch 1996a). These organizations produce and distribute pamphlets and easy-to-read books explaining legal issues. Their informational workshops provide tools for identifying violations of rights and using the law.[4] Despite this activity the obstacles to promoting legal literacy are tremendous—illiteracy, unreliable and expensive communications, poor transportation, and constraints on funds and time, and so on.

In the early 1990s, most people with whom I spoke felt that ordinary Tanzanians were unacquainted with the efforts toward law reform, which, though laudable, had little reach beyond urban areas. Nevertheless, it was important to direct attention to those groups that were working explicitly on law reform, because they were the nexus for the global/local process that shapes legal consciousness. Poised to raise the consciousness of the broader public, they acted as conduits and filters for information about the law and legal processes. Although heavily funded by donor governments and international nongovernmental organizations, they did not simply adopt global discourse wholesale. Rather, they translated its aims, making them relevant to the Tanzanian context. I examined the discourses produced by these groups as evidence of the scope and content of legal consciousness among some Tanzanians and also of the processes whereby changes in legal consciousness are stimulated.

In attending events hosted by these groups, conducting interviews with members, and reviewing their materials, I defined and redefined my position as a researcher partly in response to their understandings of me. Initially, group members assumed I was affiliated with the donors who were providing funds for legal projects, an impression I quickly dispelled. For example, my use of a laptop to collect data in court was mistakenly viewed as indicating my participation in a USAID (United States Agency for International Development) project to computerize court records. Some people were visibly and understandably disappointed when they realized that discussions with me would not yield a project grant. Their disappointment, which gave rise to feelings of inadequacy on my part, compelled me to renew my commitment to conducting research that, while not offering material benefits, would directly assist the groups in their activist goals. Given that most groups had few personnel knowledgeable about law, and very limited resources, they were especially in need of information about legal processes, which I was in the process of gathering.

When I returned to Tanzania in 1996 for some much-delayed follow-up research, I renewed contact with people working on gender issues in Dar es Salaam by attending a weekly workshop hosted by the Tanzania Gender Networking Programme. At the end of the session, the facilitator announced the talk for the next week and then turned to me. "Perhaps we could hear from you about your research. Can you present in two weeks?" I silenced the inner voices of protest that often inhibit my presenting work in progress—I had barely collected enough data, the analysis was sketchy at best, my conclusions still murky—and agreed. My acceptance was an acknowledgement of TGNP's position that foreign researchers owe a debt to the local community. I understood this to mean that if I gained background information useful to my research through attending their workshops, then the group had a right to receive something from me that would address their goals with respect to gender transformation. After accepting, I began to see the session as an opportunity to expand the participatory aspect of my research.[5] Though welcome, TGNP's invitation raised a dilemma: Could the workshop be both a site for data collection on legal consciousness among these influential Tanzanian feminists as well as an opportunity to provide them with the legal information that they requested? The dilemma bothered me until I realized that this was precisely the contradictory challenge that I had been avoiding in previous research. And, to my relief, the group itself provided a framework for conducting research on law while at the same time raising legal consciousness.

Learning from and with Feminist Activists

Begun in the early 1990s in Dar es Salaam, TGNP co-ordinates the efforts of a variety of individuals and organizations working on gender issues in the country. Their downtown office, resource library, and weekly workshops provide valuable opportunities for learning about gender and networking with gender activists and scholars. Weekly workshops hosted by TGNP involve a wide range of people—Tanzanian and foreign—with interests in gender, including consultants, basic researchers, lawyers, students, media workers, personnel in local and international NGOs, workers, and lobbyists. In the workshops information on diverse topics (e.g., cancer prevention, the education of girls, and media sexism) is shared, challenged, and applied practically. The related goals of sharing information and raising consciousness are addressed through a range of tactics. It was in this context that I began to appreciate the importance, and difficulty, of participating in feminist activist research, that is, in research on gender that blurs the boundaries between scholarship and activism.

One of TGNP's chief missions is to facilitate research, and many members conduct research projects for TGNP, academic institutions, and other organizations. TGNP's approach to research, especially their critique of methodology, is influenced by a model of participatory research developed in African contexts. As one of their founders writes:

> Participatory research emerged in the region as a critique of both liberal and marxist research, which used survey and other neo-positivist approaches uncritically. By the mid-1970s, in conjunction with efforts to decolonise social science generally, the issue of "who researched whom" began to be raised (Kassam and Mustafa 1982). The positions of the researcher and the researched became a critical issue, along with the structure of the entire research process. Sources of funds, audience, style of presentation of results and the structure of decision-making in the design and implementation of the research plan were subject to inquiry. The aim was democracy in the production of knowledge, and the empowerment of working women and men. To what extent were professional researchers capable and willing to share power and resources with villagers, farm workers, urban slum dwellers, indeed with students and research assistants? Were participatory researchers prepared to follow up on the political issues raised by participants, and remain involved when confrontation between "the stat" and "the people" developed? (Mbilinyi 1992: 63)

Participatory research had its origins in the heady days of Tanzanian socialism, when many academics were concerned with social transformation in a recently independent nation facing challenges of poverty, illiteracy, and global marginalization. Influenced by Paolo Freire and others concerned with education and social transformation, this approach demands that research be relevant to social change. There was no question that researchers should address controversial questions and through their research provide answers leading to social transformation (see also Shivji 1993). The focus on politically engaged research resembles action anthropology developed in the 1960s and 1970s by Sol Tax and others and most likely was influenced by some of the same social movements.

Despite many social and economic changes in Tanzania in the several intervening decades, especially in the move toward a market economy and privatization since the early 1990s, the orientation toward activist social research remains popular.[6] Far from being an insular approach, relevant only to East African contexts, participatory research directly engages the recent debates over science, truth, and objective observation that postmodernism has ignited, albeit from a very different perspective. Drawing on Mary Hawkesworth, Marjorie Mbilinyi of TGNP argues:

> We can make use of some of the "tools" developed in research methodology, while denying the ideology of instrumentalist science and value-free knowledge. Action-oriented critical feminist theory and practice requires knowledge about "the" world which is consistent and reliable. The procedures used to produce the knowledge we hear and read need to be reported and transparent, so readers can have a basis from which to judge the "known." The researcher/writer needs to tell us her own positions and multiple identities so readers can understand the "knower." By bringing herself and the overall research process into the terrain of analysis itself, she demystifies science and helps decentre hegemonic discourses of power and knowledge. (Mbilinyi 1992: 59)

Mbilinyi's perspective resonates with post-modernist discussions of the production of knowledge, particularly with respect to the positioning of the researcher; however, it is forged outside the preoccupation with text and authorship that characterizes the post-modernist turn in social science in the North American context. Mbilinyi and others associated with TGNP position themselves as scholar/activists, readily moving among these interrelated roles. This positioning does not imply a dismissal of basic research, theoretical language, or academic credentials. Rather, participants consistently attempt to "demystify" theoretical concepts and their own research process for those unfamiliar with academic or other technical discourses. Many of the scholar/activists with whom I spoke expressed the belief that data should be gathered and shared in order to effect social and political transformation, specifically to reduce poverty and oppression.

TGNP's commitment to sharing information, and the research process itself, is reflected in the form of their weekly workshops and described in their handbook:

> The animated workshops organised by TGNP exemplify the creation of space and time to generate and share relevant and accurate knowledge. Resources are shared during and after the workshops, and networks are created and strengthened, in order to increase women's access to the basic instruments and tools of production and reproduction— including the PRODUCTION AND REPRODUCTION OF KNOWLEDGE. (TGNP 1993: 35)

This model confirms that workshops are a site for gathering information and, at the same time, for sharing it in order to raise consciousness. From their perspective, consciousness comes through involvement in the exchange of information. At TGNP, the format of the animation workshop is altered depending on the participants involved and their needs, the issues at hand and the context, and the changing demands of transforming society. How best to produce and reproduce knowledge is problematized in the workshop itself, and remains an issue with which TGNP deals repeatedly. The production and reproduction of knowledge is central to consciousness and consciousness-raising; thus, as I came to see, these workshops provided a fruitful setting for exploring how scholar/activists engage the problem of consciousness.

Planning the Workshop

TGNP members collaborated with me to develop the format for my working session. We agreed on the title "Discussion of Recent Feminist Activism for Women's Legal Rights." In part, I hoped that calling it a "discussion" would avoid situating me as the central authority in the workshop and would encourage the active participation of attendees. In planning a workshop to meet the goals of both gathering and presenting information, and doing so while occupying a defensible role, I relied on the participatory approach that typically characterizes TGNP sessions, which is "based on animation techniques, backed up by a non-hierarchical structure of organisation led by facilitators instead of chairpersons. Participants from different locations in society, share skills and knowledge derived from years of experience in promoting women and gender issues" (TGNP 1993: 23). Animation techniques are designed to:

—raise consciousness and awareness of gender and other social issues;
—lead to action which is designed and carried out by participants themselves;
—validate the experience and knowledge of participants;
—provide a means to resolve most conflicts, through dialogue;

—increase solidarity, self-esteem, and self-confidence;
—produce many new and creative ideas;
—inspire and energise; and
—are democratic, participatory and enjoyable. (TGNP 1993: 24)

From my experience, TGNP regularly achieves these laudable goals, which makes for lively gatherings. Although they generally begin with "brief presentations of main issues.... the main content of TGNP workshops is provided by participants themselves, by means of discussions in small groups and in plenary" (TGNP 1993: 24). Animation techniques such as role play, case study, and games are important means of stimulating interactive discussion. Handouts or background documents offer participants something concrete to reflect on and to take with them, perhaps to share with others. Animation methods provide the opportunity for a wide range of people to voice their opinions and try out their ideas.[7]

I planned to begin the workshop with a brief description of my research on law and gender in Kenya and Tanzania, as TGNP members had expressed their interest in learning about theory, especially concerning law and gender. After my presentation, I planned a "practical exercise" in which everyone would work in small groups to arrive at a "judgment" on actual matrimonial cases from a Tanzanian court. Then in plenary we would discuss the results of the exercise. Since TGNP often ended such workshops by articulating a set of action-oriented goals, any remaining time would be used to assess collectively how the struggle for women's legal rights was progressing or should be progressing in the country. These components represented a balance between giving and receiving information, between data collection and active consciousness raising, and thereby reflected the group's stated ideology of what a useful workshop might be.

TGNP publicized the session. Moreover, I encouraged many of the people with whom I had conducted research to attend, believing that the workshop would provide an opportunity for them to express their views on specific cases and to also evaluate the direction of my work. Another anthropologist, my good friend Mary Porter, who was traveling with me at the time, asked with uncertainty if she should come. We puzzled over this, and once again I realized that my initial hesitancy came out of a belief that Mary's familiarity with my project might "influence" the information participants would offer. Accordingly, I welcomed her and encouraged her to participate along with the others. She helped enormously by taking notes at the session. My request to audio-tape the session was granted.

The Workshop

Getting Started
The workshop involved about twenty-eight people—Tanzanian and foreign—with interests in gender, including consultants, researchers, lawyers, students, media types, workers, and lobbyists. Most were women (about twenty) and ages ranged from late teens to late sixties, with most in the middle. A TGNP-appointed facilitator was responsible for making introductions, moving the workshop from one phase to another, keeping time, and generally "animating" the participants.[8] Her efforts

generated the exuberance of the workshop, which included singing and dancing, heated debate, challenging questions, and recriminations for behavior that was not "gender-sensitive." The facilitator's personal interest in theater led her to encourage creative participation throughout the workshop. For example, she began the workshop by getting us all on our feet in a circle. Holding hands, we greeted those near us and then each stated our name and affiliation. I noted about ten stalwart TGNP staff and members, eight students, several foreign researchers, and five others employed in government or NGOs. Among the participants was one lawyer and a quite famous magistrate, who came at my invitation since I had been observing in her court. The facilitator started a "break-the-ice" song that encouraged people to celebrate their positive personal qualities. She sang, "I want someone happy, happy like me," dancing in the middle of the circle until she was joined by another TGNP member who then chose another trait. The students and foreigners hung back, while TGNP members joyfully participated. One, who had just returned from visiting her newborn first grandchild, sang, "I want someone grandmotherly, grandmotherly like me." Everyone was surprised but pleased when the magistrate danced into the center, and we laughed when she asked for someone "serious" like her. A TGNP member responded to the call to be serious in a teasing sort of way. The song brought us all together singing and dancing or swaying, and yet it made evident the distinctions among us. Those refraining from a turn in circle's center were generally the outsiders, not TGNP members. Some people clearly identified themselves in relation to others in the group. For instance, a young woman named herself an "activist," perhaps to encourage others to be radical. Defining oneself in the community of others is an especially important part of the workshop experience, as the description below will show.

Given the technical terms and my own linguistic abilities, I delivered the talk in English. Most TGNP workshops held in the city were, at that time, conducted primarily in English. However, at some sessions people agreed to use Kiswahili or simply used it as they participated.[9] In a tone resembling the style I use in teaching undergraduates, I focused on breaking down concepts relating to law's role in social change (e.g., resistance and hegemony) and offering examples from my research in Kenya of how law shapes gender relations. My theoretical model depicted law as having the potential to reproduce gender relations or to transform them. This choice intentionally addressed my finding that many Tanzanians, including women involved in activist work, viewed law as relatively intransigent, that is, as a rather ineffectual site for social change. I was especially concerned to show that Muslim women (albeit in Kenya) were successful in using Islamic law courts to resist oppression by husbands and ex-husbands. Finally, I suggested that Tanzania and other post-colonial nations were still facing difficulties in resolving family law disputes, given the plurality of legal and normative orders (e.g., customary and religious laws) and that current discussions of women's legal rights offered the opportunity to engage some of these issues from a feminist perspective.

The workshop format defies a researcher's ability to predetermine results; unanticipated events occur. For example, immediately after my lecture, instead of moving into discussion groups, the facilitator called on the magistrate to respond. Although I realized that the facilitator was paying proper deference to this respected figure, who was one of the first Tanzanian women in the judiciary, I was concerned that a professional opinion might silence others. To my relief, the magistrate admitted that Tanzanian law was "a ball of confusion" that no one really understood, mostly because of the mixture of colonial, post-colonial, religious, and customary legal systems mentioned in my lecture. Her comments seemed to level the floor as we moved into small group discussions of family law cases.

The next section briefly describes two of the cases and interprets the group deliberations, which provide insight into legal consciousness. Moreover, the workshop, as a site for consciousness raising, also reveals the *process* of transforming legal consciousness. Situations in which legal consciousness is transformed generally involve people situated in particular social positions. Examining how participants actively negotiate their positions with each other is an important aspect of this research technique.

Deliberating and Educating

For the practical exercise, I chose three matrimonial cases from those I had collected in courts around Dar es Salaam (see Appendix A).[10] My choices were guided by an interest in obtaining reactions to some of the most vexing issues in Tanzanian matrimonial law, including laws that had been targeted for law reform or legal literacy efforts. Moreover, each case directed attention to legal processes. I prepared handouts of each case in English and in Kiswahili. Each handout described the facts of the case briefly, including quotes from the testimony of the petitioner and respondent, and concluded with a directive designed to stimulate discussion: "You are the Judge. Give your opinion on this case." This was followed by a series of questions: "(1) What is the outcome of this case? Who wins? Why? (2) What would you ask the petitioner? What would you ask the respondent? (3) When they leave the court, what advice would you give the petitioner and the respondent? (4) What does this case show about law and gender?" Another handout outlined the session (Appendix B). I participated in one group and audio-taped another, for which Mary Porter took notes.

Case One involved Amina's claim for divorce from her husband Ali. She argued that Ali ended their two-decade Islamic marriage through a letter of divorce and then abandoned her for several years without paying the post-divorce maintenance as required under Islamic law. The couple had appeared previously before the Muslim council of elders, a mandatory step before suing in civil court; however, the elders did not confirm that a divorce had taken place. Amina was now seeking a divorce certificate and maintenance from the civil court. Ali, who insisted that they had never divorced, wanted to return to the elders in order to use Islamic law.

Initially, the group discussing this case had few opinions. Most were Christians, and several said they knew little about Islamic law. One TGNP member hesitantly advanced that all marriages, whether Muslim or Christian, are subject ultimately to Tanzanian civil law. Though initially hesitant to intervene, I agreed when she looked to me for confirmation. And I added that a disputing couple must first pass through a council of elders (such as the Muslim council) before going to civil court. The group then focused on an issue of justice: Ali had abandoned Amina for a long time. Angry at him, they granted a divorce, dismissing as irrelevant his desire to return to the Muslim council.

The deliberations of this group illustrate several tendencies I observed through other forms of research. In groups of mixed religion, Tanzanians rarely discuss religious law in any detail. The same is true of mixed ethnic groups and customary law. Focusing on state law and issues of common-sense justice masks most people's lack of knowledge about religious law but, more importantly, avoids arguments over religious differences. The silence about legal pluralism resembles other tactics for preserving national unity in a population that consists of equal numbers of Christians, Muslims, and other believers (e.g., Hindu, "traditional" religion) spread among 120 ethnic groups. In public discussion, legal plurality is erased or minimized by deferring to secular state law. Consequently, knowledge of particular laws is limited, and the connection between state law and religious law is little understood, even by the highly educated and the legally trained. As I have argued elsewhere, this situation, aptly termed a "ball of confusion," means that many women do not understand law, and urban, Muslim women, in particular, fail to secure their rights through the state (Hirsch 1996b). This case highlights the difficult issue of who has authority to interpret Islamic law, which Ali raised by telling the Magistrate that he had no desire to deal with religious issues in a government court. By dismissing his claim as "irrelevant," the group kept the focus on secular law and, more importantly, on notions of just behavior.

Case Two involved the division of property between Lucy and John, who agreed to divorce after a long Christian marriage. Lucy demanded money and the matrimonial home, alleging that she had contributed to its construction, though she had no receipts. John disputed her contribution and claimed the house for himself. Appearing as a witness, their daughter supported Lucy's claim by stating that her mother used profits from selling vegetables to help build the house.

This discussion group included the magistrate previously mentioned and the only lawyer at the workshop. Together, they controlled the discussion, educating the others about matrimonial property, which they admitted was a controversial issue. They explained an important High Court decision which held that a wife's contribution to the household, including unpaid domestic labor, must be considered in property division. According to the lawyer, proving the wife's contribution poses the main difficulty in such cases, and she urged group members to keep records of financial transactions and duties performed. She and the judge admitted, however, that many lower courts did not follow the High Court ruling and that women were routinely denied property they deserved. Following the advice of those with legal training, this group decided that John and Lucy should sell the house and divide the profits. The lay members called for equal division but were persuaded by the magistrate that this would not hold up in Tanzanian courts, which tend to recognize men's contribution as larger.[11]

In both groups, participants other than those with legal training mostly asked questions or commented on what was fair or just. Their discourse resembles what Conley and O'Barr (1990) have identified as the tendency of the powerless to focus on equity or justice in relationships, rather than legal rules. However, there is a striking contrast between the Tanzanian case and findings about legal discourse in the United States. Specifically, in the workshop, few people offered comments based on personal experiences with the law (cf. Merry 1990; Yngvesson 1994). No one expressed a sense of entitlement to use the law. Rather, they voiced grave doubts about the fairness and accessibility of the legal system. Distrust of and contempt for

Tanzanian legal institutions had been consistently expressed throughout my research as ordinary people in Dar es Salaam, and also legal personnel, cited the law's corruption, inefficiency, and lack of resources. For those in the workshop, going to court was not viewed as a means to a just solution. It is important to note that not all Tanzanians share this view of law. As Sally Falk Moore (1989) has shown, many people in rural areas have significant experience, some positive, with customary law, local councils of elders, and rural primary courts. But the workshop participants—mostly middle-class, urban women—generally had little connection to courts in part because of their doubts about law's efficacy.

Sharing Results
As the groups finished deliberating, the facilitator brought everyone together for a plenary discussion of the "rulings" on the cases. Representatives from each group made elaborate presentations, writing their answers to the questions on flip charts that were given to me afterwards. Those presenting answered each question in turn, paying special attention to providing advice to the disputants. I included this question in order to assess how people would respond to the opportunity to raise legal consciousness. Group One advised Amina to "Please take care of your life." And they expressed their intolerance of Ali by telling him to "Please mind your own business" and accept the court's ruling. Their advice to Ali reveals a paradox: Even people who are skeptical about whether law helps to achieve justice can insist that courts, as authoritative institutions, be obeyed. Group Two—called "the legal minds" by other participants—argued that advice should be given *before* the judgment in order to encourage a settlement. In answering the question of what the case meant for gender relations and law, the lawyer, as group representative, spoke eloquently about Tanzanian women's vulnerability with respect to proving monetary contributions. She parodied couples who "sit together and say: 'Oh this is our property. Oh these are our things. It's *all* for the children.' But this is how it should end up, with a straight face: 'Uh uh dear, I love you, but we have to put it on paper.'" The participants laughed at her scenario and clapped enthusiastically when she ended by calling on everyone from lawyers and judges to the community at large to be "sensitized to ... have a wider definition of what is matrimonial property." The facilitator ended this part of the workshop by jokingly proposing a new clause in the marital vows: "As we become one, we should also document."

At the end of the presentations, I explained that the findings of each group reflected what had actually transpired in court. Participants were pleased that they had come up with similar answers; however, some were surprised that the courts had actually gotten it "right." By learning that at least some women receive support and perhaps even justice in Tanzanian courts, workshop participants experienced a shift in legal consciousness, given the skepticism about law as expressed during the deliberations. The use of real cases from Tanzania, rather than hypotheticals or ones from another context, probably made this shift more likely. I added that, while the decisions in these cases were not unusual, few such cases made it all the way to court. Had the groups decided differently than the courts, we would also, no doubt, have had an interesting conversation resulting in effects on legal consciousness.

The workshop offered other opportunities for raising legal consciousness. The lawyer, judge, and facilitator were quite overt in their efforts to raise legal consciousness by providing specific information about legal processes and urging participants to improve their legal literacy. Although their efforts addressed the workshop goal of raising legal consciousness, they also created some tensions between the gender activists trained in law and the majority of workshop participants who, similar to most middle-class Tanzanians, have little direct connection to law. For example, when the presenter for Group One announced that they had decided for the petitioner "with costs," the magistrate let out a hoot of involuntary laughter. She apologized, explaining, "You simply can't award costs in a case like this." Group members were a bit chagrined at having been reminded that the law is a domain requiring expert knowledge. This exchange called into question my assumption that raising legal consciousness should be primarily about empowering individuals—especially those not trained in the law—to take an active role in learning about and questioning law. The prominent roles played by the legal personnel perhaps worked against individual empowerment by reminding some participants of their dependence on judges and lawyers for information. This aspect of the intragroup dynamics had the benefit for me of disrupting any notion that I was the sole "expert" in the group. More generally, the presence of any experts in an activist group necessarily raises political questions of who is empowered to act on particular matters and who directs the group's actions. As these brief moments in the workshop suggested, such questions, which are the subject of intragroup struggle, are necessarily studied as part of the process whereby legal consciousness is shaped.

Building Consciousness and Relationships

As the workshop turned to the last topic—the future of law and gender in Tanzania—a TGNP leader apologized for rising to address another issue: the way in which the few male participants had monopolized the discussion. She was particularly distressed that two of the three group presentations had been made by men. The lack of "gender balance" in the workshop—many more women than men—had been a concern from the beginning. In forming the initial discussion groups, the facilitator had insisted that each group have at least one man and had assigned them accordingly. In criticizing the prominent role played by the men, the TGNP leader said, "This workshop is also about us. It is about our group and our struggle to work together politically." For their part, the men protested that no one else had volunteered to present. The leader insisted that, in the future, the men must refuse and thereby force the women, who were looking very guilty by this time, to speak for themselves.

Those involved in participatory workshops and other TGNP events (e.g., meetings, conferences) are always negotiating their relationships with one another. In TGNP and similar groups, such negotiation includes conscious efforts to interact with each other in ways that not only facilitate specific activist goals but also instantiate values inherent in the group's ideology. By directing attention to aspects of interaction—such as who should speak and for how long—TGNP members try to avoid further empowering those already vested with structural power (e.g., men,

elders, educated people). The instance described above is a relatively rare overt polic-
ing of these understandings, which are likely more often the target of self-censorship.
As a matter of theory that relates to the analysis of legal consciousness, understandings about how language should be used are aspects of the context in which speech
about law is produced (see Hirsch 1998). These "metalinguistic" aspects of the context
must be taken into account when evaluating what a particular instance of speech
reflects about the legal consciousness of the speaker. Depending on the context, those
who are knowledgeable about law might refrain from speaking for reasons relating to
these interactional prerogatives. In this workshop, the usual TGNP understandings
about language and power were not followed, as men and "experts" monopolized
discussion. No doubt in TGNP's very next workshop, participants would be more
mindful of the discursive rules.

After the discussion of interaction, there was only time for the facilitator to direct
everyone's attention to a flip chart on which she had just written: "Should bad laws
be reformed and accommodated? Or should they be destroyed?" The audience
answered with a resounding "destroyed," and the workshop was closed on a note of
high energy. As a plan for action, the call for destroying bad laws failed to capture
the complexity of the preceding discussion. However, these final moments were
another reminder that consciousness raising happens in particular settings where
participants' goals, including the regulation of interaction and group solidarity building, are often paramount. For TGNP members, the goal of invigorating the group
was an important aspect of what this workshop on law, or indeed any workshop, was
supposed to accomplish.

The Feminist Working Session as Method

My workshop experience suggests that studying deliberate attempts to alter legal
consciousness, including those planned by the researcher, can illuminate the content
of consciousness and the process of legal change. At the same time, much of what I
learned about legal consciousness through the workshop confirmed findings based
on other research methods. The value of the working session as a method in legal
anthropology lies, at least in part, in the opportunities it offers for raising legal
consciousness and for altering the relationships of those involved in the research
process. To assess the working session as a method for legal anthropology, I reflect
below on three issues raised at the outset of this chapter: the methodological difficulties of studying legal consciousness, the relationship between researcher and
others, and the blurred boundaries between scholarship and activism in sociolegal
studies.

Participatory research does not overcome the central methodological difficulties
of studying legal consciousness. Legal consciousness, in the sense of an individual's
understandings of law, is studied through observation and analysis of non-linguistic
and linguistic behavior (e.g., interviews, court testimony). On the one hand, the
workshop is simply another context for the expression of understandings, with all of
the complexities that researchers analyzing those understandings routinely confront,
including prevarication and misspeech, misinterpretation, and the danger of creating
a static model of consciousness, which, in reality, is always changing. On the other

hand, the workshop format provides a context in which expressions of understandings are usually defended and explained, thus affording a rich text for analysis. Moreover, in contrast to interviews, the metadiscourse about consciousness is produced through interaction among a variety of workshop participants, an important variation on the speaker/researcher model. The workshop experience reminded me that all linguistic data must be evaluated in light of the explicit and implicit discursive rules operating in a context. Finally, an interactive, participatory approach often stimulates changes in consciousness; by sharing understandings, participants come to think differently about cases, laws, and legal personnel. Such moments of change—so difficult to capture through other research methods—are illuminating for the study of legal consciousness.

During the working session the participants, including myself, disrupted the standard roles of researcher and research subject that typify other methods (e.g., observation in court, interviews with judges). For example, as a researcher I was not always positioned as an authority gathering data. Rather, I treated the other participants as active subjects in the research process not only by allowing them to control the interaction but also by explicitly adopting their format for gathering and sharing knowledge. Throughout this chapter I have acknowledged that the use of participatory methods took the discussion in directions that precluded some of my research goals; however, by allowing other participants to pursue their individual and collective goals, I believe I enhanced the experience of the research process for everyone.

In struggling to define my role prior to the workshop, I confronted my own assumption that legal consciousness could and should be investigated without the "influence" of the researcher on those being studied. Because it was not possible to sustain this view while meeting the participatory expectations of TGNP members, I came to view consciousness as fluid and contextually shaped in ways that preclude quasi-experimental measurement. Thus, my analysis accepts, and also interrogates, my role as someone actively involved in changing consciousness through my research methods and, indeed, my work more generally. Moreover, changes in my own legal consciousness (e.g., that I began to think differently about documentation) and feminist consciousness are important aspects of the research experience.

The last point above returns to the issue of the relation between scholarship and activism. The session allowed me to pursue my political goals of providing Tanzanians with information about law at the same time as I continued my research. Moreover, my participation allowed me to interact with Tanzanians as both an activist and a scholar. By "activist," I mean someone who intentionally attempts to effect social change. As I learned more about legal consciousness in Tanzania, I came to see that, whether or not researchers are activists, our scholarship can be used for political ends, with or without our consent. Specifically, many of the campaigns for law reform, democracy, and women's rights in Tanzania, which are primarily funded by donor nations and international organizations, routinely rely on information produced by scholars of law and society, women's studies, anthropology, and other social and policy sciences. The scholarship underlying these projects, which circulate globally, is most likely to be produced by scholars in the United States and Europe rather than in Tanzania or on the African continent, a fact that reflects the economic and educational marginalization of the African region. Thus, as a scholar, I participate (often unwittingly) in an unequal circulation of information that has an impact on the legal consciousness of Tanzanians. By situating myself as an activist, I make my participation in the production and distribution more intentional and thus more directly reflective of the feminist politics I support. Moreover, I can explicitly address the interests and goals of Tanzanian researchers and activists, who are striving to guarantee that law reform reflects local circumstances as well as donor prerogatives. Given the difficulties they have in pursuing their scholarly and activist goals—e.g., lack of funding, donor constraints, and sexism—my efforts might provide additional assistance; I consider this assistance an appropriate feminist goal.

Conclusion

Most significantly, the workshop described herein deepened connections among the other research participants and myself by providing a context for the fusion of multiple roles: feminist, activist, and researcher. I learned a good deal studying legal consciousness from within those fused roles. Following Deborah Gordon (1995), I reject the notion—raised in recent critical literatures—that the errors of Western feminist anthropologists disqualify us from pursuing research outside our cultural homes: "Precisely because so much Western feminism has enacted the colonial logic of either saving non-Western women from themselves or their men or turning them into curious, quaint relics through government or missionary work, applied or advocacy anthropology needs not to be abandoned but to be reconstructed." In reflecting on her own use of experimental ethnography in promoting women's literacy, Gordon (1995: 374) writes that "[b]y 1991, action-oriented research seemed academically dated." Indeed, by then, the debates over law, policy, and advocacy in law and society scholarship had long since cooled, and, even during their heyday in the seventies, anthropologists were only marginally involved. Although legal anthropologists have consistently exposed the operation of power through our ethnographic work, we have been less concerned with reflecting on the power dynamics of research itself. The few scholars who have begun to explore the potential for activist academic work in legal anthropology offer the possibility of making the subdiscipline less vulnerable to charges of irrelevance, callousness, and neo-colonial objectification (see, e.g., Coutin, in press). Gordon (1995: 384) appropriately concludes that participatory, action-oriented anthropology "needs to be rethought for feminist purposes." Efforts to rethink and reconstruct participatory research should be undertaken in the name of feminist legal anthropology, given the possibilities for social change lying within law and also legal anthropology's longstanding interests in power and gender. Developing interactive, politicized research that raises legal consciousness and simultaneously studies it will not only expand feminist causes in relation to law but might also invigorate legal anthropology from a feminist perspective.

Appendix A

Case One

Petitioner: Lucy; Respondent: John

Summary from a case file in Kisutu Resident Magistrate's Court, Dar es Salaam, Tanzania:

Lucy and John were married by Christian rites in 1975. At the beginning of 1991, they started to fight, and Lucy left the house. In 1992, they went to the Marriage Reconciliation Board. They agreed to divorce, but John did not agree to give Lucy the house, which was newly built. Lucy came to Resident Magistrate's Court and asked for a divorce and that house and some other money.

During the case the following occurred:

In front of the judge Lucy said: "We started to fight a little while ago. Sometimes John hit me when he was angry. We agreed to divorce in front of the elders. Now he doesn't want to give me my rights, like the house and money. It's true he worked hard and got a good salary. But I worked too. I did all the work at home and also I grew vegetables, which I sold. I used the money to buy our food at the house."

In front of the judge John said: "The trouble started when she ran away from the house and refused to return. I went after her to her parents. We went to the local leader and the CCM and the police and the Marriage Reconciliation Board. Finally, we agreed to get a divorce. But I cannot agree that she gets the house. That house I built it myself. She did not help with anything."

In front of the judge John and Lucy's child (twelve years old) said: "Mama got money from her business of selling vegetables. I think that sometimes she paid for things at home."

Case Two

Petitioner: Amina; Respondent: Ali
Summary from a case file at Kibaha Primary Court, Coast Region, Tanzania:

Amina and Ali were married under Islamic law in 1968. They have lived in separate houses since 1989. In 1992 Amina came to Resident Magistrate's Court. She claims for a divorce, maintenance for three months, and a house.

During the hearing, the following occurred:

In front of the magistrate Amina said: "My husband gave me a letter of divorce in 1989 and he threw me out of the house. I stayed for three months of 'edda' and he didn't give me edda maintenance or even a divorce certificate. I accused him at the Council of Muslims and he told me that he had never divorced me. The elders of the Council said that there was no divorce. They agreed with him. I stayed apart from him for three years and he didn't give me anything. Now I want a divorce certificate and edda maintenance and the house."

In front of the Magistrate Ali said: "It's true that I gave her letter of divorce in 1989. Because she started to go out of the house inappropriately, and she went with other men. I certainly didn't throw her out of the house. She left herself after getting the letter of divorce. Now I don't want to divorce her. I want to go back to the Council of Muslims again. I married my wife under Islam and I have no need to come here to a government court. Better that matters like this one are discussed at the Council of Muslims who will use Islamic religion."

Amina answered him: "I don't go out of the house inappropriately. I want to be divorced."

Appendix B

TANZANIA GENDER NETWORKING PROGRAMME (TGNP)
GENDER & DEVELOPMENT SEMINAR SERIES
SEMINA KUHUSU JINSIA NA MAENDELEO

August 7, 1996 3:00–5:00 PM.
DISCUSSION OF RECENT FEMINIST ACTIVISM FOR WOMEN'S LEGAL RIGHTS
Facilitated by:
Susan F. Hirsch
Department of Anthropology
Wesleyan University
Middletown, CT 06459 USA

I. Introduction: Theoretical Approach to Gender and Law in Society
    A. Law as a Site for Hegemony and Resistance
        1. In relation to the state
        2. In relation to gender hierarchy
    B. Challenges of Struggling for Women's Rights in Plural Legal Systems
        1. Example from Kenya
        2. Potential Issues for Tanzania

II. Practical Exercise: Three Matrimonial Cases in Tanzanian Courts

III. Discussion: Strategies toward Women's Legal Rights in Tanzania
    1. What should be the relation between religious law, customary law, and secular law?
    2. Should a nation with many groups have one legal code for every area of the law? For example, family law? Inheritance law? Assault law? Land law?
    3. What is the best way to achieve legal rights for women? Law reform? Better application of laws by police, lawyers or judges? More informed use of law by citizens? Efforts outside the legal system? Other approaches?

Notes

1. Sally Merry (1990) defines legal consciousness as the ways people understand and use the law.
2. Under Tanzania's Marriage Act of 1971, people can initiate claims at any level. The Primary court is generally used by those with limited means and no access to representation (see DuBow 1973).
3. A brief initial research trip to Dar es Salaam in 1992 convinced me that some scholars, legal practitioners, and feminist activists were interested in law as a tool for social change, and they lamented that the topic was relatively unstudied in Tanzania. My consultations with an admittedly small number of urban, educated Tanzanians was a first step in incorporating the views of local people into my research agenda. Many anthropologists mask this type of initial "participatory" research contact in their ethnographic writing, perhaps sensing that it falls short of really connecting with "the people" being researched. However, for many of us, such connections with the elite facilitate and shape our research. Moreover, the legal elite in most nations has significant influence in determining legal consciousness as well as the general role of law in society. As scholars, we may wish to make their formative contributions a more explicit aspect of our discussions of method.
4. TAMWA (Tanzania Media Women's Association) also operates a Crisis Center that serves as a shelter for battered women who have left their homes. Several groups offer limited counseling and assistance in taking claims to the police and to court. Efforts to increase legal literacy are undertaken primarily in the capital; however, teams of women from various organizations have traveled to other regions to lead interactive workshops. Several radio and television programs address the issues. These programs include Maoni Yangu (My Opinion), Women's Half Hour, and Jarida ya Wanawake (Women's Journal).

5. As Susan Coutin wondered in conducting participatory research on sanctuary activists in the United States: "Was I studying myself?" (Coutin and Hirsch 1998: 11).
6. For background on the political and economic transformation of Tanzania as it relates to gender, see Tripp 1997.
7. Writing about Alice Walker's fiction, in which characters converse in multiple voices and perspectives to depict varying experiences that lead ultimately to a collective understanding of oppression, Faye V. Harrison deftly phrases the sense shared by TGNP members that many different kinds of people can contribute usefully to the production of knowledge: "Intellect and knowledge, then, are not elitist and exclusive; they are based on a collective, historicized consciousness, in which the experience and wisdom of the folk can invoke authority in negotiating the resolutions that the younger, formally trained intellectuals make in their thinking and in their lives" (Harrison 1995: 238).
8. At TGNP and similar groups, the facilitator plays a significant role in workshops. The facilitator guides discussion in a participatory manner and organize participants for role play or other activities:

   During workshops, facilitators are constantly in action and on the move; writing flip charts and putting them on the wall ...; making suggestions to one another; presenting case studies and singing songs; listening to participants and observing the general mood; changing the agenda and time-table when necessary. FLEXIBILITY is the hallmark of animation workshops—participants are often surprised, at first, by the openness in the workshop organisation, and the changes which occur in time-tables. Participatory management styles are different from what people are used to in formal education settings and the workplace. With appropriate guidance, however, most participants "catch up" with the mode of operation, and become active promoters of animation techniques. (TGNP 1993: 10)

   My facilitator was reassuring and poised to animate those reluctant to participate. Her strong presence created a dynamic which meant that attention focused on me only for a small portion of the workshop (i.e., during my talk).
9. Kiswahili is the national language of Tanzania. Most Tanzanians speak it fluently, although it is not the first language for many. English is spoken as a third language by some. Kiswahili is the medium in primary schools, and English is used for secondary and higher education. At TGNP, facilitators informally poll those present to determine which language to use, and group members routinely translate for those unable to understand Kiswahili or English.
10. Under Tanzanian law, matrimonial claims can be initiated at any judicial level (e.g., Primary Court or a higher Magistrate's Court); however, parties must already have sought a hearing with a community-based mediation panel recognized by the state.
11. Although Tanzanian case law recognizes the contribution of spouses to matrimonial property, including through unpaid labor (e.g., childcare), this controversial finding is not widely known or accepted, even by some judicial personnel. One lawyer told me that he believed the decision was "discriminatory" because it took property from men.

References

Bell, Diane, Pat Caplan and Wazir Karim
    1993   Gendered Fields: Women, Men and Ethnography. New York: Routledge.
Conley, John and William M. O'Barr
    1990   Rules versus Relationships: The Ethnography of Legal Discourse. Chicago: University of Chicago Press.

Coutin, Susan and Susan F. Hirsch
    1998   Naming Resistance: Ethnographers, Dissidents, and States. *Anthropological Quarterly* 71.1: 1–17.
DuBow, Frederick
    1973   Justice for People: Law and Politics in the Lower Courts of Tanzania. Ph.D. Dissertation. University of California at Berkeley.
Gluck, Sherna Berger and Daphne Patai eds.
    1991   Women's Words: The Feminist Practice of Oral History. New York: Routledge.
Gordon, Deborah
    1995   Border Work: Feminist Ethnography and the Dissemination of Literacy. *In* Women Writing Culture. R. Behar and D. Gordon (ed.). Pp. 373–389. Berkeley: University of California Press.
Harding, Sandra
    1987   Feminism and Methodology. Bloomington: Indiana University Press.
Harrison, Faye
    1995   Writing against the Grain: Cultural Politics of Difference in the Work of Alice Walker. In Women Writing Culture. R. Behar and D. Gordon (ed.). Pp. 233–245. Berkeley: University of California Press.
Hirsch, Susan F.
    1998   Pronouncing and Persevering: The Gendered Discourses of Disputing in an African Islamic Court. Chicago: The University of Chicago Press.
    1996a   Productions of Battering in African Contexts: Local Examples in Global Processes. Paper presented at LSA Annual Meeting, Glasgow.
    1996b   The Feminization of Islamic Courts: Varieties of State Intervention in East Africa. Paper presented at AAA Annual Meeting, San Francisco.
    1994   Kadhi's Courts as Complex Sites of Resistance. *In* Contested States: Law, Hegemony, and Resistance. M. Lazarus-Black and S. F. Hirsch (ed.). New York: Routledge.
Jackson, Jean
    1986   On Trying to be an Amazon. *In* Self, Sex, and Gender in Cross-Cultural Fieldwork. T. L. Whitehead and L. Price, ed. Urbana and Chicago: UIP.
Mbilinyi, Marjorie
    1992   Research Methodologies in Gender Issues. *In* Gender in Southern Africa: Conceptual and Theoretical Issues. R. Meena (ed.). Pp. 31–70. Harare: SAPES.
Merry, Sally
    1990   Getting Justice and Getting Even: Legal Consciousness Among Working-Class Americans. Chicago: University of Chicago Press.
Minh-ha, Trinh T.
    1989   Woman, Native, Other: Writing Postcoloniality and Feminism. Bloomington: Indiana University Press.
Mraz, Jacqueline
    1997   Of Law and the Tears of Things: Notes on the Varieties of Legal Consciousness. PoLAR: Political and Legal Anthropology Review. 20.2: 101–114.
Reinharz, Shulamit
    1992   Feminist Methods in Social Research. New York: Oxford University Press.
Shivji, Issa
    1993   Intellectuals at the Hill: Essays and Talks 1969–1993. Dar es Salaam: Dar es Salaam University Press.
Strathern, Marilyn
    1987   An Awkward Relationship: The Case of Feminism and Anthropology. Signs 12.2: 276–292.
TGNP (Tanzania Gender Networking Programme)
    1992   Gender Profile of Tanzania. Dar es Salaam: TGNP.

Wilson, Lynn B.
    1995   Speaking to Power: Gender and Politics in the Western Pacific. New York: Routledge.
Yngvesson, Barbara
    1993   Virtuous Citizens, Disruptive Subjects: Order and Complaint in a New England Court. New York: Routledge.

Chapter 2
Trekking Processual Planes beyond the Rule of Law

Philip C. Parnell

Fantastik, Cinch, Ultra Spic and Span, Lysol Antibacterial Hand Gel, Shiny Sinks Plus, and Ultra Dawn—all are bottled-up allies stored under my kitchen sink for battle against the dangers of germs. Most often my antibacterial arsenal is out of sight and mind, cloistered behind two cabinet doors, and unlinked from activities and goals that are compartmentalized and uncoupled from cleaning the kitchen sink, swabbing the linoleum floor (with built-in shine), and assuring house guests that the enemies of order and offenders of the senses are under control. Compartmentalization, separating and confining different life activities so that the debris of one does not cascade into and disrupt what is going on in another, is a way of getting by while being effective in some relationships and wreaking havoc in others. In doing battle against living a life that could be charted through hydraulic equations, compartmentalization cognitively cages the serpent of ignorance—bringing him under control in one endeavor is not going to make him pop up in another—and disguises as mere cracks social fault lines that stretch across a life, making them useful separations of the fragments of a fractured existence. This is moral survival. Through compartmentalization, my Ultra Dawn cannot link me to incipient cultural streams of eradication that may be snaking through a subconscious collectivity of my culture. The "damned spot" can grow without becoming inconvenient. Through the mentalistic segregation of identities, the body that conquers germs at home is not a body that would organize society along notions of contamination.

While experiencing ethnography, unexpected questions can appear in those compartments that segment an ethnographer's life. Field experiences may intimate new interpretations for everyday actions that have meanings either unexamined or assumed by the ethnographer at home. While doing ethnography, I look for how people put together the disparate components of their lives by constructing coherent pathways of meaning from one action to the next through practices that render the bricolage of complex culture, culled from communications of the past and present, a recipe for shared goals. Disputing can play important roles in gathering the recipe's components and tasting their combinations. Others' theories, the inventions of scholars, can perform this function, once the data is in tow, but one of the ethnographer's tasks is to discover the theoretical inventions within nonacademic systems of thought and action.

Searching for the synapses of difference among others can reveal the ethnographer's own mental/physical disconnections, especially when disconnections and compartments are components of the schema of convenience—getting by—rather than explanatory devices. The researcher can discover the political accommodations of home—concepts and beliefs that accrue intellectual concessions—when they appear as boundaries in the pursuit of ethnographic knowledge. As I will attempt to illustrate below, collected intellectual concessions can signal "no" to the body, even when the mind says "yes." I will interrogate such boundaries of accommodation and look at how some of the compartments and schema of convenience have been and are being challenged by the questions rather than answers of ethnographic research.

This is, then, a discussion of being on the boundaries of legally constructed worlds while conducting ethnography, and of what happens when those boundaries are crossed. It is an exploration of the marginalizing forces of thought and law—of how law, in crossing cultures, can act as a "modulation," limiting the ethnographer "like a self-deforming cast that will continuously change from one moment to the other" (Deleuze 1992: 4), and of how law can enclose and render unknowable some sites of habitation and work as well as the people within them. To do this, I draw on portions of an ongoing ethnography within three realms of law: law in the United States as it is communicated through media, roles of law in the United States as they are integrated into the researcher's maps, and roles of law in metropolitan Manila.

Law's Body, Ethnography's Mind

In the summer of 1987, at the beginning of my ethnographic project in the Philippines, I had crossed my first legal boundary conceptually, but my body backed away. My goal was to study disputing across culturally different groups in metropolitan Manila in the effervescent months and years that followed the People Power Revolution of 1986. In prior research in Mexico, I had been intrigued by the unexpected formation of regional and urban–rural intergroup coalitions in my field site in the highlands of Oaxaca, where I lived for a total of two years during 1973–1974 and 1984 (Parnell 1989). These coalitions transformed processes of interpersonal disputing, and the roles of state-level courts in highland villages, which had been the initial focus of my research. I wondered if the resurgence of democracy in the Philippines was similarly transforming intergroup relations from the Marcos period of martial law and dictatorship. I was also interested in how intergroup disputes fit into the formation of Manila's rapidly proliferating coalitions—the colossus of Philippine socializing and networking had been awakening from a long period of suppression. By 1987, some exiles from the Marcos years had returned to Manila, and the democratic underground had burst through the thinning surface of fear to form proliferating political coalitions and nongovernmental organizations.

I had concluded that a squatter settlement could be an ideal site for my research. Migrants from different islands and cultural regions, who shared Tagalog as a first or second language, had been gathering for decades in settlements of the poor, and they were not always sorted out territorially on the basis of cultural history. But during the first two weeks after my arrival in Manila, while riding through the streets of the city in ubiquitous and joyfully decorated open-air elongated jeeps (jeepneys) and gazing at settlements of the poor, I began to struggle with "common sense." Some large shanty settlements appeared impenetrable and unfathomable, their entrances and exits invisible through patchwork shanties, each appearing to lean in the direction of disintegration on a neighboring house. Other settlements were small groupings along polluted drainage ditches and urban rivers, their inhabitants bathing in water that carried human debris.

The urban poor of the United States are enclosed by the cultural construction of danger. Though I once lived in one of the poorest sections of New York City (the monthly rent on a one-bedroom apartment was $25.00), the last 25 years of my knowledge of urban American poverty has been infiltrated by the media-based narratives of the United States legal system constructed in the times of drugs, AIDS, the diminution of the civil rights movements, interethnic struggles, escalating punitiveness, criminal cartels, and the politics of racial fear. So I wondered while contemplating a field site what would be my chances of surviving while living in poverty as it has been depicted in the United States. The popular image of the urban poor has been constructed through American law as one of moral decay and desperate random violence. In law-constructed electric images, territories of the poor only approximate safety when their landlords are courts and cops. They are not even safe when under constant governmental surveillance. The poor are corralled in neighborhoods, such as those in Los Angeles described by Mike Davis, through eagle-eyed police helicopters and border barricades that allow only residents to enter and exit: The poor neighborhood exists as a "foreign land" (1990). Desjarlais captures the images:

> A common problematic vision is apparent in many accounts [of the homeless poor in the United States]: the homeless live in an underworld; they are a ghostly, animal-like brood who threaten the peaceful, artful air of cafes, libraries, and public squares. Television shows and Hollywood movies like *The Fisher King* create pictures of ghostly, ragged vagabonds haunting the post-industrial wastelands of American cities, while newspaper accounts thrive on images of death, transgression, and grotesque bodies. Distinct themes or images are often paired: a morgue in a museum, a runaway in a library, blood and Beethoven. The homeless themselves serve to counter images of health, wealth, purity, and high culture. The imagery passes swiftly, unquestionably, as if it was in the nature of sentences, or putting words onto paper, to set pain against beauty and wretchedness against form. (1997: 2)

As I traveled some of the poverty-lined streets of Manila the critical academic in me waged a mental battle with familiar images of the poor that took form in U.S. media-based reporting on law and crime. While moving at traffic jam speed past the lives of the urban poor, I fought a growing sense of personal isolation. My mind and body were at odds. As an anthropologist I believed that popular images of poverty as dangerous and morally decaying had no essential connection to the behavior of the poor. I knew that societal forces do not necessarily constitute moral and immoral bodies through allocation of just desserts. For some, the poor offend and become dysfunctional within a consumerist view of the future increasingly pervasive in currents of United States theme park culture in which safety and security may be achieved more effectively through being a successful consumer than through the forces of official law (Project on Disney 1995). The urban poor are relegated to the terrains of state-controlled public space. This message was strong in Metropolitan Manila, where pervasive poverty viewed on and from the streets vanished behind the doors of enclosed multistory mega-malls guarded by uniformed private security forces armed with repeating rifles. The malls, eerie in their stony acoustics, brightly lit, with antiseptic corners, were citadels in the battle against disorder (as signaled by the markings of human debris).

The strongest internal battle I experienced was the one waged by the ethnographer against the compliant citizen. In analyses of the urban human landscape, geographer Edward Soja contends that forces of governmental control construct spatial boundaries among populations, mostly economic classes, in order to place people in the geographical spaces where they are most accessible to the goals of capitalist enterprises (1989). For example, law and its enforcement, over time, create urban patterns of population distribution in which blue collar workers live mostly in areas near the industrial enterprises that need to employ them. Governmental enforcement of economic boundaries expressed as urban spaces can be related, analytically, to the social isolation of worker populations from other economic groups, which inhibits their movement up social and economic ladders—a cumulative ghettoization of workforces through police actions and their effects on popular perceptions. I sensed in these first weeks in Manila that the social and cultural boundaries the media communicate while using law as a context for dramatizing lives and intergroup relations in the United States had been inscribed on me and, as I experienced their control, on my body (Rose and Miller 1992). I could not physically move across those boundaries, even though they had been constructed in another land and even though the intelligence I had gained through anthropology had frequently led me to question and meditate on the well-traveled social and cultural canyons of American legal culture. I could gaze at and consider the squatter settlement, but I could not enter it. I was confined to an isolating world of state-empowered discourse (Foucault 1977).

Experiencing these contradictions, I evaluated the squatter settlement as a field site, using law, religion, and old-time social anthropology as my yardsticks. I balanced rights and duties, their content derived in part from a Methodist upbringing in which the "lifestyle" is "service" (a motto emblazoned in red on the publication of the church in which I was reared). I recalled the Western anthropological principle, certainly of Calvinist origin, that claimed that essentialized relationships between economic conditions and culture are inevitably severed for the researcher by exposure to the inventive forces of social relationships. During my first two weeks in Manila, these reflections were condensed in the feeling that I had earned professional freedom from dangerous field work while conducting research in Mexico, as if that experience had immunized me against the symbolic forces of my own culture. This thought reinforced my sense of isolation in Manila—the feeling that I was wandering in a silenced crowd.

I sought alternatives to a squatter settlement through conducting interviews with leaders of political coalitions and reform movements (for a few days, while escaping from my conjoined feelings of humiliation and self-preservation, I formulated a project to study the Free Farmers Federation while traveling for a year throughout the Philippine countryside); but Philippine knowledgeables to whom I explained my research repeatedly directed me back into urban plights. Following their instructions, I arrived at the office of Sister Mary Paul Durazno, who was dean of a small Catholic college in Metropolitan Manila and leader of Gabriella, an umbrella organization for many of the Philippines' women-oriented nongovernmental organizations (NGOs). At the end of our interview, Sister Mary Paul invited me to join her for the inauguration of a day-care center in the poor region known as Tondo. Because some of Tondo's residents built their homes within the large smoking mounds of trash that they scavenged for recyclable goods, Tondo was a must-see stop along the routes of poverty tourism—a symbol for Western eyes of oriental desperation, or the failures of colonialism.

We met the next day away from the trash heaps in a quiet region of two-to-three-story concrete shops. The community we visited wound through alleyways lined on each side by homes covered with sheets of tin. They converged on a larger home at the end of their shared passageway, one in which the owners lived with the pigs they were raising to vend. I had been warned about the smell, and, predictably, pigs in Manila, to the unsophisticated nose, have much the same odor as those in Indiana. I wondered how long it would take for the odor to disappear with familiarity, just as the everyday smell of antibacterials can become reassuring.

The day-care center had been funded by the government of New Zealand; we joined the cultural attaché to the New Zealand embassy and his modest entourage. The focal point of the center's inauguration was a performance by children of the neighborhood. As the guests became an audience we bunched together on very short stools to watch the children perform a piece of guerilla theater, the first I had witnessed outside of New York City. This performance was not diluted for paying audiences, the actions and movements of the actors identical to short clips of Chinese anti-U.S. guerilla theater—often involving violent actions against symbols of the United States—that I recalled from childhood newscasts. I had studied Tagalog, so I knew that the young children stomping as their staccato arm movements pointed and repainted wooden rifles were, rather than a welcoming committee, local interpreters of the politics of foreign aid. I looked around wondering if the New Zealand benefactors, all of whom appeared stone faced, knew what was going on. It was not unreasonable (for me at that time) to wonder if this community was aligned with the New People's Army (NPA), the armed wing of the indigenous Philippine communist party that had been waging a battle against the official Philippine government, mostly in the countryside. Respected Manila newspapers proclaimed red alerts almost daily, announcing the possibility that the NPA could be mounting attacks within Manila. Uniformed United States soldiers had been shot recently on the streets of Manila. The presence of United States military bases not far from Manila was opposed by many Filipinos, and their opposition was inflamed by sexually abusive conduct attributed to American military personnel. A little more than a week prior to this visit, renegade soldiers of the Philippine armed forces had led an armed attack on the presidential palace in an attempt to overthrow the government of President Aquino. I took the messages of the children seriously as expressions of anti-imperialist sentiments held within this alleyway squatter neighborhood.

This encounter in Tondo located me as an outsider in the settlement, and my reaction was to spontaneously challenge this undeniably accurate identity. As if by the power of suggestion, I became a resister to both my role as audience and the way it reaffirmed the marginalizing forces of state law. I was in Bourdieu's world, where all action is linked to capitalist social formations (Bourdieu 1972, as characterized by Karp 1986); but what I felt was a rejection of the analytic conventionality of the choreographed settlement encounter. For the first time since my arrival in Manila, I felt at home. Rather than experiencing fear, I sensed mild elation, as if I had stepped into a cleverly camouflaged utopia in the alleys of Manila. Though far from an urban dreamscape, this squatter settlement was an escape from the clamoring silence of the Manila I had experienced while traveling from office to office through crowded streets in disrepair, breathing air as fresh as a puff on a Salem and steamed with the essence of burning lard. It was an escape from isolating encounters with the tall concrete walls of the wealthy while I walked past them along crumbling neighborhood sidewalks. The vibrant trees, bright gardens, and capital-intensive mansions they protected seemed as far from my research goals in Manila as such houses are from my life in the United States. My elation was not an epiphany, I reasoned, for I do not believe in epiphanies that capture a precise extract of society through a cauldron of cultural variables conjugating within the mind (Leslie White 1949). People—and the societies they create—are too complicated to mistake epiphanies for ethnography.

I mistakenly recognized this intellectually and politically out-of-place feeling as intuition, "the direct experience of things as they are" (Davis-Floyd and Arvidson 1997: ix). In my self-dialogue of making an ethnographic decision, the visit to Tondo evoked feelings similar to those I had experienced while living in the small villages of the Oaxacan highlands. I associated these feelings, in a romantic way, with the presence of community. In this reasoning, I had developed something of a sixth sense, one that could tell when community, as a social body, was present. As I struggled with the isolating perceptions of danger evoked by using law as an urban map (as if the city could have been divided into regions of control), identification of my feelings as those of community was centering: It gave me a traditional ethnographic place of the kind that I knew. I ignored the other cues of my visit and decided to trust my ethnographically-tuned good feelings. I could live in a squatter settlement; my past work as an anthropologist had prepared me for this research.

In retrospect, what had happened was less knowing but far more revealing than intuition. Through the ethnographic process that places the researcher's body within other-cultural settings, I had engaged in what Wikan, in translating Balinese ways of knowing, refers to as "feel-think" (1991). In opposition to the analytic and pigeonholing frameworks of Western law, in which the body and mind are separated as subjects of discipline and knowledge (Devine 1996), my body had joined with my mind. I physically crossed boundaries that the deterrence-based law of a "danger culture" (Rip 1991) had skillfully constructed. What the ethnographer had believed was the elation of finding a home among the poor, the intuited sense of community, was not intuition; rather, it was the experience of freedom, of liberation from the rule of law, not by breaking the law, but by breaking free from the deeply burrowed inscriptions on my mind and body of pervasive American legal culture. I was in a place that I had feared, and witnessed a basis for rational fear; yet, rather than being afraid, I felt at home. The body had opened to the critical mind.

The decision I made that afternoon in Tondo changed my life and has infiltrated it with a lasting strategy of approaching borders that are the haunting manifestations of ethnography's unanswered questions. Later that year, after living several months in a crowded squatter settlement, I took my first trip to Manila Bay. Standing before the open stretch of water, I felt my body expand. That experience was similar to my unfolding in Tondo as I communicated with another horizon. The doors to Manila had opened.

Living in Disputes

This liberating physical and cultural transgression weakened other boundaries and led to the challenging of additional symbolic anchors of my identities and beliefs. Following several interviews, I met with a diminutive woman with a warm smile in the open sanctuary of Kristong Hari (Christ the King), Catholic hub of Commonwealth, a sprawling squatter settlement that surrounds the Philippine Congress building (the Batasan Pambasa). Although I knew little about life among the approximately 100,000 residents of Commonwealth—such as whether or not it was a violent place—I was seeking permission to live there. My companion in the Kristong Hari pew was testing my Tagalog, the politically-favored language of the settlement. She was a Commonwealth squatter and leader of its most powerful residents association, Sama Sama. After I passed her test (an unusually appropriate rite of passage for an academic), she offered to locate a house where I could live in the settlement.

The house was a small one-room scrap wood and fiber-board shanty with a bent tin roof located in an Ilocano family compound. It was not much larger than a medium-sized dining table. Inside, the ground was broken concrete and dirt. The shanty was inhabited by small bugs and large spiders. A light bulb dangled from the roof by a wire. I learned that to conduct my research on disputes among the squatter groups of Commonwealth I would have to live in this least accommodating of the region's houses, one that was nevertheless believed appropriate for a single male. I was told that some nuclear families lived in houses no larger than mine. Next to the shanty was a rusty deep well, the location's most attractive feature. We placed a rag at the spout to filter out the rust for drinking water. Not far away, at the edge of the settlement, were two of Metropolitan Manila's largest garbage dumps. During the dry season, in Manila's constant heat, winds filled the air with particles of the dump's dry debris.

Although only three weeks earlier I had not been a resister to the all-American crusade for safety (or the control of danger), my body no longer balked at the visual signs of danger that were abundant in Commonwealth. I willfully abandoned comfort and medical safety. After I had begun to shed some of the bricolage of United States law by physically transgressing one of its boundaries, the seams that held together outward symbols of my own identity (my persona as expressed in space and time) began to unravel. Moving into the lives of Commonwealth residents had become a path rather than a passage and, lying in my shanty, I felt, as does Mary Douglas, that "There is no such thing as absolute dirt, it exists in the eye of the beholder" (1966: 12). The residents of Commonwealth lived there illegally.

Philippine Marines could have arrived any day (and in some areas did arrive) to demolish their houses. No zoning or sanitation codes were enforced there and, possibly, Commonwealth was a place where denizens of Manila came home to sleep. Under martial law, and later during my residence there, Commonwealth was swept and "sanitized" by forces of state control, not in the enforcement of health codes but toward the control of political opposition. Anonymous dead bodies in black plastic bags, looking like large podded bugs, were dropped off there by equally anonymous killers.

As I began to find my place in Commonwealth I placed trust in the idea of community tied to place. My untested assumption was that society, including ones as disparate as those in Commonwealth, could breed safety in the absence of state enforced law. But I would soon discover that this was not a "place" as I had conceptualized it. My new expectation of safety, like the former one of danger, was grounded in a false assumption, one derived once again from the role of law in the United States in the regulation of demographic patterns. Safety and danger are basic components of urban maps; but as I settled into Commonwealth they had become, for me, little more than amorphous concepts. To replace them and begin the process of reconstruction, which is ethnography, I turned to time and, as I had done with my law-based map, I relinquished my control over time. To do this, initially, I gave my time to the women of Sama Sama. This was a fortunate decision since, as I would discover, their calendar-based schedules were renegotiated prior to most of their meetings and appointments—dates and times were loose pebbles in the running streams of their lives in the settlement. In handing my time to neighbors I also gave them, for many hours of the day, physical control over my body, a further extension of my residence in the Commonwealth shanty. This attempt to move into society as it was experienced by its creators across time and space (Durkheim 1965; Sorokin 1941; Zerubavel 1981; Coronil 1997) completed a symbolic inversion of my status that had begun only weeks earlier. The consumerist male carrier of Western (and for the Philippines, colonial) law was now controlled to a large extent by poor Asian women who, in several ways, lived outside of and challenged official state law. The women of Sama Sama seemed agreeable to this arrangement—perhaps they had invented it.

Most mornings when I left my shanty for the dirt streets of Commonwealth (the shanty was too hot to be in during the day), the participants of Sama Sama and I found each other, and I went wherever they went. Their daily lives were in large part governed by the trajectories of intergroup disputes and negotiations, and soon my calendar became the same as the processes through which their disputes developed. Rarely did I know from day to day the places where I would go or the cultural boundaries that I would have to cross. With members of Sama Sama, I traveled across Manila in jeepneys and buses to municipal and federal offices and to meetings with international developers, cabinet-level bureaucrats, and some of the nation's religious leaders. Having escaped the walls of the city's sidewalks by moving into Commonwealth, I was on the other side of those walls as a guest of the urban poor. These were places of power and privilege of the sort I had rarely entered in my own society.

The members of Sama Sama were attempting to remove some of the larger roles of official law from Commonwealth and Payatas Estate, the larger urban region of which Commonwealth is a part. Their goal was that a portion of Commonwealth be treated differently from many other areas of the city through becoming a semi-autonomous region governed by a partnership between Sama Sama—representing its residents—and representatives of federal and municipal agencies. Essentially, much of local and federal civil law was to be suspended in Sama Sama's plan, and official structures of local-level governance were to be complemented by Sama Sama as a "people's organization." This was more or less the state of affairs as members of Sama Sama negotiated the future of Commonwealth and its residents with federal agencies and, at the same time, offered help to residents in the management of local level interpersonal and land-based disputes. Although legal titles to the land of Commonwealth were held by the federal government and privately, Sama Sama sought ways to nullify those titles as charters for development.

Seeking to withdraw from the state through partnership with agents and agencies of the state was a pattern rather than the exception among the squatter organizations of Commonwealth and Payatas. Originators of this pattern among the urban poor appeared to be land syndicates that controlled both large regions of Payatas land—to which the syndicates did not hold title—and the squatters who settled on it. The syndicates were trying to nullify existing law through court cases in which syndicate leaders argued they were the rightful heirs of those who held Spanish titles to what is known today as Metropolitan Manila at the end of the Spanish-American War (these titles were issued in the 1890s). The United States, when making the Philippines its colony at that time, recognized the Spanish titles. Other types of squatter organizations developed specific partnerships with state and municipal bureaucracies. These partnerships sometimes included land title cases within the squatter organizations supported governmental litigants by researching the history of challenged titles issued by the Bureau of Lands.

Sama Sama's conduit to and influence within federal bureaucracies was the Catholic church. Land syndicates operated through lawyers and corporations that were also linked to people who held positions in governmental bureaucracies. Other organizations had multiple links to religious organizations, lawyers, municipal employees, workers in land-related bureaucracies, politicians, and court employees, which they organized to reach into the amorphous world of state-related influence and power. In Commonwealth and Payatas, the power of the state was realized primarily through demolitions conducted by the police and Philippine Marines. As goals of autonomy were pursued by Sama Sama through the church and federal bureaucracies and by land syndicates through litigation, daily efforts were extended by their members to prevent the demolition of their homes and relocation to the edges of the city where industries were struggling to survive.

Though the vertical organization of rights-construction was common to societies where patron–client relations expressed important organizational principles (Roniger and Gunes-Ayata 1994), I was surprised by the absence of official governmental participation in the governance of an urban region as large as Payatas. As I looked up for the state and its law, as I was accustomed to do, something seemed to be missing. But as I moved through and with the disputes of the urban poor, I found the state beside us, at the bargaining table, at meetings in Commonwealth, in gatherings of religious groups, and in the offices of state bureaucracies. I expected to see the state through its uniformed representatives, along with representatives of welfare bureaucracies, acting beside the very poor who lived illegally on the land. But the police (and military) were here in the settlement primarily as residents, and the small amount of money for welfare was apportioned by the poor.

A battle for Payatas was going on within and across the vertical intergroup networks that reached from Commonwealth and Payatas into other Manila sites of power. Real estate corporations were also vying for pieces of Commonwealth and Payatas and had created middle class suburban housing developments there even when their titles to the land were legally uncertain. As I examined court cases concerning the land of Payatas, it appeared the formal law was playing only a small role in the creation of the future of this "city within a city." The state was present primarily through its participation in disputes among Commonwealth groups. As I followed the disputes, I located the Philippine state in social context and *as* context. My life, calendar, and map became the processes of disputing among vertical slices of the Philippine nation. Through these disputes the state, law, and community were constructed as components in the lives of Filipinos.

If I had been living outside of Payatas, basing my research solely on interviews and documents and not engaging in participant observation, I would have been meeting with the gatekeepers of information and arguing the worth of my research to nego-tiate wall by wall access to each governmental bureaucracy and its records—The Housing and Urban Development Coordinating Counsel, the Presidential Commission on the Urban Poor, the Home Insurance Guarantee Corporation, vari-ous offices of the Quezon City municipal government, and the Department of Public Works and Highways. Instead, I was finding access to information and communication the Filipino way, as part of a group that follows the well-worn paths of vertical network travel in a patron–client society through open doors. As I observed meetings that produced decisions and records, I experienced the processes for which those meetings were the punctuation and the discourse that would result in political and material changes within Payatas.

My silence was as important to trekking these networks and the disputes that acti-vated them as my abandonment of Western time, comfort, and medical safety. My opinions were received by the members of Sama Sama with disregard. I could observe but not contribute, communicate what I saw but not interpret it according to my own standards. In playing this role, which Sama Sama must have considered in some ways beneficial, I learned the role of the government in its relationship to urban poor organizations such as Sama Sama, whose members were motivated in part by the knowledge that governmental attempts to provide land and housing for the urban poor had most often failed (Parnell 1992). Private citizens and groups who had taken matters into their own hands had been the primary benefactors of the poor. Sama Sama's mantra was that only the urban poor had been able to successfully construct the infrastructures and provide the services necessary to the survival of urban poor communities and entrance of the poor into more rewarding participa-tion in the national economy. Around Commonwealth, "Herr Professor," as my teacher Laura Nader often referred to the imperialist academic in the field, would wait a long time for the uptown bus to arrive.

Living in disputes bifurcates the academic world. In Manila, where the Western models I had attempted to abandon can easily mismeasure, where I lived virtually homeless during the day and like the poorest of settlement residents at night, I found and observed practices of the Philippine political and religious elite, including some who had captured the populist imagination in the reconstruction of Philippine democracy. In contrast, my Western friends there, who also happened to be academics doing research in Manila, often evaluated their observations through the calculus of global politics and capital and Western law, measuring the Philippines as if it were a highly centralized federal political system. Some of their mental mazeways were, unintentionally, those of colonization and social evolution.

Some of my colleagues in Manila saw Philippine governmental, bureaucratic, and economic systems that had failed and could only fail when measured against the background of Western bureaucratic blueprints. By such measures, Philippine institutions would always fail, for the Philippine state I experienced as I wound my way daily from the squatter settlement through hierarchies of church and federal governance was distinctively non-Western—it worked well in ways that Western governmental institutions often prohibit. The lenses of Western time, status, and law, and notions of bureaucratic rationality, clouded the windows of Philippine institutions to those who stood outside them, even as those institutions mimicked Western calculations. Philippine processes, including disputes, governed by local time and prestations, created a governmental system that was distinctively indigenous and elusive to the devotee of straight lines and right angles. Through its leaning homes pieced together from eclectic materials, and its infrastructure that grew slowly but surely over time rather than overnight, the squatter settlement that appeared dangerous as mapped through American law had come to symbolize, for me, the vitality and creativity of the Philippine state.

Mind versus Motion

Early on one morning of field research I was standing in the street that ran by the family compound that surrounded my shanty. I had just returned from the home of a retired Philippine professor, Helen Mendoza, where I went on an occasional weekend to sleep and stave off the demons of exhaustion that accompanied constant involvement in disputes and the sustained contradictions of always-partial cultural knowledge. I had learned that exercising the option of exit was sometimes necessary in the kind of ethnographic research I was conducting. Living purposefully without asserting control can fill the mind with contradictions that one does not often encounter along consciously chosen paths; and living with the contradictions, rather than trying to resolve them with social and cultural theory, had become a productive and energizing tension midway through my research—I had decided to let Philippine time manage them. But that morning a member of Sama Sama found me and said there had been a demolition attempt in nearby Sitio Kumunoy. We were to go there.

The lore of squatter groups is filled with tales of resistance. During a demolition attempt in 1983, at a region of Commonwealth known as Gilarmi, women of Sama Sama raised a barricade and faced Philippine Marines. Even though some women were shot, members of Sama Sama returned to the barricades the next day and the Marines retreated. Victory in this dispute forged some of the conviction that constituted the foundation of Sama Sama as a group.

We traveled down the street from my home in Commonwealth and rode motorized tricycles through a middle class subdivision of nicely finished two-story homes and up a wooded hillock to the site of Sitio Kumunoy, where a few shanties had been destroyed. The squatters who were there to protest the demolition and resist the unmanned bulldozer nearby claimed, as did other urban poor organizations, that the subdivision had an illegitimate title to the land, one derived from a counterfeit "mother title" that had been manufactured by a corrupt Philippine political family.

While studying politics in the field, I had cautiously avoid being political; but in this squatter action I was concerned that if I were seen as part of a protest against government-backed legal action my right to conduct research in the Philippines would be in danger. The women of Sama Sama were protected in part by their political positioning in Philippine society. As unarmed women aligned with respected Catholic organizations, they were unlikely targets of violence at the hands of Philippine politicians struggling for approval in post-revolutionary Manila. They also had a reputation of risking their lives to protect their homes and the futures of the families who lived in them. They had sought power within governmental agencies through nongovernmental channels. Unlike men who populated more conventional political routes, they could not lose their political and economic positions through opposing the opponents of their influential allies within governmental bureaucratic institutions—such opposition, and the threat of it, was part of their strength. As a male who had arrived in the Philippines through conventional channels, I feared that I could be expelled for doing the unconventional. Although my presence in the heart of this protest had not been a conscious choice, but merely a stop along the daily routes of Sama Sama activity, the excuse "I gave them my time" could have hastened rather than forestalled any trouble that lay ahead for me.

For the first time since my arrival in Manila—when I feared living in an unknown squatter settlement—I sensed danger and wanted to flee the dangerous place that Kumunoy seemed to be; but the signs of danger there were as false as the faux danger created for poverty and the poor within messages conveyed through the idioms of American law. While the resisters sang songs and ate sticky rice spread across banana leaves, and I conducted a cursory survey of the small Kumunoy settlement, other members of Sama Sama and their allies were organizing forums throughout Manila for the construction of change. The action of Sama Sama with residents of Kumunoy, the threat of confrontation that arose from our presence, existed for the creation of organizational linkages and alliances within Kumunoy and, more importantly, across Payatas and within governmental bureaucracies. Moving away from the hillock, I accompanied Sama Sama members to meet with a regional homeowners (squatter) association. We then went to a large meeting (of about 25) at the Quezon City municipal building with representatives of the Presidential Commission for the Urban Poor, the subdivision developers, and the Catholic allies of Sama Sama. After that, we attended an even larger meeting at the Home Insurance Guarantee Corporation, also attended by residents of Kumunoy, where the Corporation's commissioner announced he would suspend all demolition (and construction) in areas such as Kumunoy (vast regions of the urban poor declared Areas for Priority Development under President Marcos). Other meetings immediately followed—one with representatives of land syndicates; another with leaders in the Housing and

Urban Development Coordinating Council; and, finally, we went to a larger convocation of homeowners associations in the region of Payatas who began to explore linkages to Sama Sama, its allies, and philosophies. From the action involving a few homes in Sitio Kumunoy, the larger reaction to stall misuses of the law, organized governmental force, and land titles spread throughout intergroup networks that extended from Kumunoy to Sama Sama and into Metropolitan Manila's land and housing intergroup networks and political associations. The organization of power that was spun off the demolition attempt and its resistance arose in so many places that I could not have begun to document all of them, much less visit them and watch the society that their participants envisioned take form.

As I stood on the hillock of Sitio Kumunoy, apprehensive with the buzz of each approaching tricycle, the dispute that I had envisioned provoked a desire to abandon the confrontation—to flee. I had looked out over the subdivision and listened through its eerie silence for approaching police, Philippine marines, and bulldozers. As the dispute unfolded in my mind, I was standing with citizens in confrontation with the state. At that moment of apprehension, my Western map of the future and the state gained control of my consciousness. Though the unarmed protest was peaceful, the members of Sama Sama and its allies were asserting control over legally titled land. I surveyed the horizon for forces of the state without realizing that I was standing with those forces. I was in the midst of democratic, not illegal, action. The resistance quickly led to the official state's suspension of the right of developers, who held legal title to the land, to demolish and construct new structures within Sitio Kumunoy. Once again, official law had not been invoked to define the situation.

I realized that the role of law in my constructions of the future was a barrier to my understanding of what the present and future were like to those among whom I was conducting ethnography. This also clouded the meanings of their choices as they engaged in disputes. The members of Sama Sama were not imagining or living in a Western state or legal culture. The state that they were putting together, from the settlement to the Presidential palace, was not one that I could have imagined at that point in time. I would have to travel through it many times over before I could begin to sense and articulate its nature. I had lived in cities, as the subject of states, and as a member of a nation. I had read about state systems as they had been captured, constructed, and assumed by researchers. But I could not imagine the state in which I was living—the Western state was still unfolding in my mind.

In the months that followed, as I moved through the processes of the Philippine state as they arose within the actions of members of Sama Sama, land syndicates, their opponents, and official governmental policy makers, as well as in the shifting alliances and oppositions across the squatter networks and vertical alliances that governed Payatas, I began to see the state around me rather than above me. I saw it in the actions of my neighbors rather than the immanent contexts of my own culture. I realized that the neighborhoods of Commonwealth were communities only through their attachments to the people who lived and worked behind the concrete walls of office buildings and wealthy neighborhoods. They became communities as their residents spent hot days crisscrossing Manila on network spinning jeepneys and on buses that clogged the business districts of Manila with social traffic jams. With this realization, I gazed from the jeepney at the squatter settlements we passed at traffic jam speed, and

I no longer sensed isolation and danger. Members of the settlement communities were in the Jeepney with me.

Conclusion: The In and Out of Legal Ethnography

Pervasive legal culture and the importance of the rule of law may be a barrier to ethnographic research on law for Western researchers in other societies as well as in their own. Law is so intricately intertwined with the search for safety, control, profits, and longevity that it is an important force in ways we organize the value of life. Ethnographers, through their presence, often intervene from the outside in the lives of others; and, for Western ethnographers, law relates particular ways of intervening to the pursuit of valued goals. Given the role of law in the mapping of Western lives, traversing mentalistic boundaries constructed with the help of law may be as important to gaining an ethnographic stance as locating the questions of ethnography in the contexts of space and time.

Practicing the ethnography of law has the potential to marginalize researchers who are from rule of law societies as it reveals law as a non-essential endowment that can transform many kinds of relationships by imbuing them with threatening qualities. Such dangers, through the exercise of law over time, can appear intrinsic to relationships. Getting to know the legal Other through ethnography can reveal the legal map of problems and their resolutions as sometimes less a guide to human survival than a plan for societal control. The high value placed on the rule of law, increasingly a quid pro quo for countries seeking to enlarge their minor roles in international marketplaces, can render struggling to survive for some groups an unacceptable excuse rather than a right. Just as the good cleanliness of my kitchen in Indiana may disguise the health consequences of not developing the ability to live with ever-encroaching natural worlds, can law that is seen as so intrinsically good that it can immunize social life from problems diminish the ability to see, respond to, and live with cultural differences? Can law as cleanliness become a compulsion rather than a constructive tool for survival amidst the endurable human ability to differ? Hierarchies are intrinsic to law, and as environments are organized over time through law's manifest priorities of safety and control, qualities such as freedom and ambiguity (uncertainty) can become the bedfellows of danger. But now I am constructing an ethnographer who stands outside the law in a rule of law society, one who must float in a pervasive legal culture and obey the law while fearing the rising legal tide. Is compartmentalization—confining law's force cognitively—the only way out of this life-fracturing dilemma?

Members of Sama Sama used their marginality to figure out and reconfigure processes and policies for the distribution of urban land in Manila. They created rights by moving outside the official law onto democratic ground where they forged a partnership with the state. Similarly, from the marginal locations of ethnography's questions, the ethnographer of law can formulate both investigative agendas and conceptual frameworks that further the goals of research and place law in its many larger societal contexts, rather than allow these agendas and contexts to be established for the researcher by the institutions of official law. By "moving out," the ethnographer can "move in" with something to offer—an understanding of the legal that is not on the map of pervasive legal culture.

References

Bourdieu, Pierre
    1972   *Outline of a Theory of Practice.* Cambridge: Cambridge University Press.
Coronil, Fernando
    1977   *The Magical State: Nature, Money, and Modernity in Venezuela.* Chicago: University of Chicago Press.
Davis, Mike
    1990   *City of Quartz: Excavating the Future in Los Angeles.* New York: Verso.
Davis-Floyd, Robbie and P. Sven Arvidson
    1997   *Intuition—The Inside Story.* New York: Routledge.
Deleuze, Gilles
    1992   "Postscript on the Societies of Control." In *October.* 59, Winter: 3–7.
Desjarlais, Robert
    1997   *Shelter Blues—Sanity and Selfhood Among the Homeless.* Philadelphia: University of Pennsylvania Press.
Devine, John
    1996   *Maximum Security—The Culture of Violence in Inner-City Schools.* Chicago: The University of Chicago Press.
Douglas, Mary
    1966   *Purity and Danger: An Analysis of Concepts of Pollution and Taboo.* London: Routledge and Kegan Paul.
Durkheim, Emile
    1965   *The Elementary Forms of the Religious Life.* New York: Free Press.
Foucault, Michel
    1977   *Discipline and Punish: The Birth of the Prison.* Translated by Alan Sheridan. New York: Vintage Books.
Karp, Ivan
    1986   "Agency and Social Theory: A Review of Anthony Giddens." *American Ethnologist* 13(1): 131–137.
The Project on Disney
    1995   *Inside the Mouse—Work and Play at Disney World.* Durham, NC: Duke University Press.
Parnell, Philip C.
    1989   *Escalating Disputes—Social Participation and Change in the Oaxacan Highlands.* Tucson, AZ: University of Arizona Press.
    1992   "Time and Irony in Philippine Social Movements." In Joann Martin and Carolyn Nordstrom (eds.) *The Paths to Domination, Resistance, and Terror.* Berkeley: University of California Press.
Rip, Arie
    1991   "The Danger Culture of Industrial Society." In Kasperson, Roger E. and Peter Jan M. Stallen (eds.) *Communicating Risks to the Public—International Perspectives.* London: Kluwer Academic Publishers: 345–365.
Roniger, Luis, and Ayse Gunes-Ayata
    1994   *Democracy, Clientism, and Civil Society.* Boulder, CO: Lynne Rienner Publishers
Rose, Nikolas and Peter Miller
    1992   "Political power beyond the State: problematics of government." In *British Journal of Sociology* 43, 2: 173–205.
Soja, Edward W.
    1989   *Postmodern Geographies: The Reassertion of Space in Critical Social Theory.* New York: Verso.
Sorokin, Pitirim A.
    1941   *Social and Cultural Dynamics,* vol. 4. New York: Bedminster.

White, Leslie
   1949   *The Science of Culture—A Study of Man and Civilization.* New York: Farrar,
   Straus.
Wikan, Unni
   1991   "Toward an Experience—Near Anthropology." In *Cultural Anthropology* 6, 3:
   285–305.
Zerubavel, Eviatar
   1981   *Hidden Rhythms—Schedules and Calendars in Social Life.* Chicago: The
   University of Chicago Press.

CHAPTER 3

LEGAL ETHNOGRAPHY IN AN ERA OF
GLOBALIZATION: THE ARRIVAL OF
WESTERN HUMAN RIGHTS DISCOURSE TO
RURAL BOLIVIA

Mark Goodale

Introduction

In this chapter I explore the effects of globalization on legal ethnographic fieldwork through an examination of the impact of the arrival of Western human rights discourse to rural Bolivia during the last ten years. Beginning in the late-1980s and continuing through the 1990s, several events in Bolivia coincided that would form the foundation for this development. First, there was a national debate in Bolivia during the mid- to late 1980s over the upcoming 500 years observations in 1992. This debate was accompanied by the formation of new indigenous rights groups and the strengthening of existing organizations with progressive or radical tendencies, particularly the influential labor unions. The impact of the new movement—framed now in terms of indigenous rights and largely united, something that is unusual for Bolivian social movements—was most dramatically represented by the turbulent 1990 march by indigenous rights groups from Trinidad to La Paz, an event that captivated the nation and forced a national dialogue about the marchers' demands, which were broad in scope but centered around claims that traditional authority structures should be given legal effect at the national level, and that rural lands should be protected from the encroachment by large landowners and corporations, especially in the Bolivian Amazon.

Probably the most striking and lasting effect of the resulting debate over the march and related events was the widespread acceptance—even among groups, like the growing middle class, that had been traditionally ambivalent[1]—of one of the debate's central premises: that there were distinct groups of Bolivians who should be defined as "indigenous" Bolivians, and that these groups had special rights as a result of this status. The working definition of "indigenous" and the rights that were felt by many people in Bolivia to accompany this status, were not understood ambiguously or organically in most cases; rather, both were derived from specific international charters or proclamations of relevant United Nations working bodies, like the International Labor Organization. The best example of this direct link is Convention 169, a broad statement of principles created by the International Labor Organization in 1989. In 1991, Bolivia became one of the earliest of only fourteen countries to ratify the convention.[2] Moreover, the package of progressive reforms developed by the Sánchez de Lozada government during the early and mid-1990s—especially Popular Participation and the Law of Educational Reform—was a specific attempt to put the ideas of international charters like Convention 169 into practice.[3]

Second, the late 1980s and 1990s saw a massive influx of nongovernmental organizations (NGOs) to rural Bolivia. Although nongovernmental organizations like the Rockefeller Foundation had turned their attention toward Latin America as early as the Green Revolution of the 1960s (see Cueto 1994), NGOs intent on helping local populations "develop" did not turn their full attention toward rural Bolivia until the late 1980s. Many of the most active NGOs in the region, especially in the mid- to late 1990s, consciously reformulated the approach of earlier organizations in that they purported to adopt "indigenous knowledge" regimes into their local planning strategies.[4] This was indeed a striking departure from earlier waves of NGO activity, which were typically seen by local people in rural Bolivia as patronizing, heavy-handed, and, at times, misdirected. Beginning in the mid-1990s, NGOs became more self-consciously collaborative by hiring local leaders to participate in decision-making over resource allocation.[5] Further, development activities were often preceded by initial meetings between NGO workers and local leaders that were intended to place the intentions of the organization within a specific human rights context as defined in international charters.[6]

Background of Research Area

Sacaca is in many ways a typical town in the north of Bolivia's Potosí Department. Created in the 1570s, during Viceroy Toledo's infamous "reductions"—in which thousands of Indians were forced, or "reduced," into the Spanish-style towns that are found throughout the Andes—it is the capital of the province Alonso de Ibañez. The 1992 Bolivian census lists the population of the town of Sacaca itself at about 2,000 people; the remainder of the province's 21,000 people live in the almost 200 hamlets spread over the province (Instituto Nacional de Estadísticas 1992). Sacaca is one of the only towns in the province that has regular motorized transportation to a major city—Oruro—and it features electricity (since the early 1980s), potable water, and, as of the early 1990s, television. Because Sacaca, like the rest of the region, is at such a high altitude (about 3620 meters), crop production is limited to the high altitude cultivars like potatoes and quinoa, with crops like corn being grown in a few of the province's lower intermontane valleys. Apart from agriculture, townspeople, like everyone in the province, have an assortment of animals that they pasture, including llamas, cattle, sheep, goats, burros, and pigs.

But outside of Sacaca, where the ayllus[7] are predominant and people live in hamlets[8] of varying sizes,[9] the picture is different. The hamlets lie at varying distances from Sacaca, and the distances are measured in "leagues," a league being understood in the region as the distance that a healthy adult can walk in one hour (it works out to about five kilometers). The closest hamlets are at about one league from Sacaca and the farthest are at about fifteen leagues or more. Throughout the province the soil is generally poor because of both the high altitude and overgrazing by Old World animals (who tend to pull plants up by their roots, as opposed to llamas, who eat the stems and leave the plant intact); periods of micro division of land also impoverish the soil because of the pressure to plant in fields that have not lain fallow long enough.

The Arrival of Western Human Rights Discourse to Alonso de Ibañez

The arrival of NGOs to Alonso de Ibañez created a new and important dynamic: the coupling of ideas that explicitly validated indigenous knowledge and lifeways with a power structure that was clearly alternative and superior to the government in terms of technological sophistication and access to resources. Further, the arrival of NGOs gave people in the hamlets a way to bypass traditional networks of power and political influence. The clearest example of this is the way in which union leaders began advising ayllu authorities to end important traditional practices. But the NGOs have also contributed to something of a shift in worldview among hamlets that have interacted with them. Local people, especially local intellectuals who have worked with NGOs, have come to view themselves within a much different context, one that contrasts sharply from what came before: a model of the world reflected by a social hierarchy that was seen to be permanent and which existed to their clear disadvantage.

The final event that formed the foundation for the arrival of Western human rights discourse in Alonso de Ibañez, and its transformative effect on local constructions of legality, was the social revolution created by the series of progressive legislation that I referred to above. The package of legislation, which involved amendments to the national constitution, was created in a political atmosphere in which the rights of Bolivia's indigenous peoples—again, as outlined in specific international charters—were given a prominent place. Although several components of this legislation have come under sustained critique in Bolivia for the way they express contradictory messages,[10] most of the central themes—decentralization, local power over decision-making and resource allocation, legal recognition of indigenous social structures (e.g., ayllus), and educational reform—have won widespread support from indigenous rights groups and commentators alike. In fact, the major debates in Bolivia during 1998–1999 among indigenous rights leaders and progressive intellectuals centered not around the validity of these core themes, but rather whether the new Banzer government would prove an obstacle to continued efforts to realize them in practice (Ricardo Calla, personal communication, 1999).[11]

For many legal intellectuals in Alonso de Ibañez, both Popular Participation and Educational Reform in particular have been lightening rods for a kind of social activism not seen before in the region. In reflecting a vision of cultural autonomy expressed in terms of human rights, Popular Participation—the centerpiece of the progressive legislation of the mid-1990s—and related legislation give hamlets and aggregations of hamlets in the province a new and forceful way to maneuver strategically vis-à-vis the dominant interests in the provincial capital. What makes the Popular Participation movement so influential in the province—and effective in the eyes of many local intellectuals—is the fact that it does not function merely as a practical solution to rural problems as understood in urban centers; rather, its emphasis on indigenous lifeways, and its radical adoption of decentralization as an ordering principle, give it a measure of symbolic importance that connects the national and the local in a way rarely seen in Bolivia. The result of this reformulation of the symbolic, combined with the appearance of NGOs discussed above, has been the rise of a boldness in action and sophistication in understanding by hamlet intellectuals that has taken many officials in Sacaca by surprise.

This chapter will explore the implications for legal ethnographic work of these dynamics from two angles. First, in the next section I will examine the ways in which Western human rights discourse has transformed local constructions of legality through one case study: a description and analysis of the Servicio Legal Integral (SLI) that was active in the province between 1995 and 1998. The SLI was created in order to provide a legal mechanism in the province for the protection of the rights of women and children, rights which were understood as subsets of the broader doctrine of human rights. And second, I will then focus on the work of the SLI's director, Lucio Montesinos, because he has served as an important link between transnational legality and local legal affairs in Alonso de Ibañez.

The final section of this chapter will discuss the methodological problems related to the research on the penetration of Western human rights discourse in rural Bolivia. I will show how the legal ethnographer can respond to the challenges created by globalization by adopting, among other techniques, both a multisited approach to research, and a focus on individual legal-intellectual biography. The central problem for the legal ethnographer in these circumstances is how to track the global–national–regional–local articulations associated with globalization by re-envisioning the research site as multileveled and multidimensional.

A. Case Study
Sacaca's Servicio Legal Integral, 1995–1998   The Servicio Legal Integral ("Legal Services Center"; hereinafter "SLI" or "Center") that existed in Sacaca between 1995 and 1998 was officially known as the Centro de Servicio Legal Integral de Sacaca. The SLI was authorized by the Ministerio de Desarrollo Humano, Subsecretaría de Asuntos de Género (SAG).[12] The Claretian Church in Sacaca made the formal application to La Paz and when the approval was given, they were designated as the managing agency. They also provided initial funding for the Center, along with UNICEF and Sacaca's mayor. The Claretians continued as official managers of the Center until April 1998, when they withdrew their support for the Center; the management then passed to the mayor until August 1998, when the SLI was closed. While it was open it had a staff of two: Lucio Montesinos, the town's only titled lawyer, who served as director; and an assistant who moved to Sacaca from La Paz. She was a young university graduate in forestry who had an interest in human rights and specifically the rights of women and children. She had many duties at the Center—which was located in a two-story house off Sacaca's plaza—including processing new arrivals, giving advice to and counseling to women who decided to stay at the Center, and in general, serving as a house mother to the group that was living at the Center at any one time. She was also responsible for maintaining records related to the Center's activities.

Montesinos worked with the assistant during interviews of new arrivals. He was responsible for all legal functions of the Center, including giving legal advice to women who came to the Center for help, making decisions regarding the validity of cases at intake interviews, planning legal strategies for cases that were deemed worthy of prosecution, and, finally, prosecuting cases as they made their way through the system. Montesinos had traveled to La Paz to receive extensive training in human rights and family law both prior to and during the operation of the SLI. The Center only handled cases that fell within the Center's mandate; normal criminal or civil cases were not processed, but cases in which men were plaintiffs *were* accepted. In some time periods they filed complaints in surprisingly high numbers, although they were discouraged from doing so by the Center's staff. Further, during the three years of its existence, very few women from the provincial capital itself utilized the Center's resources; the overwhelming majority of plaintiffs were women from the hamlets.

Besides serving as a legal resource center for women and children in the province, the SLI was also a refuge for women who were fleeing abusive home environments. The impact of the Center on women in the province is difficult to overestimate or fully appreciate. Although physical and psychological violence by men against women and children is an omnipresent feature of life in the province, the reaction to the Center proves that such violence is not "accepted" as some essential feature of local culture or, even more, as a reflection of an essential "Andean culture." Rather, given the ability to formally protest such violence and the conditions that create it, and the opportunity to take concrete steps to better their situation, women in the province will react quickly and unequivocally, despite the disruption in family life their reactions cause.

Apart from its practical impact, it is also not easy to fully appreciate the symbolic importance of the Center to women in the province. Women in the province had certainly never considered the fact that they had formally recognized legal rights *as women,* not to mention the possibility that a state-sanctioned entity existed to safe-guard these rights on their behalf. Beginning in the 1970s, women in many of Bolivia's mining centers became politically active within the context of the growing international women's movement (see Barrrios de Chungara 1978), but there has never been any indication that this movement had an impact in Alonso de Ibañez. In general, women are largely excluded from taking active official political or legal roles. This is not to say that women do not exert influence unofficially, or that they are not active legal actors; but the lack of formal opportunities to serve in authority positions has meant that women have not had the same exposure to wider political and legal ideas as men have, particularly men like the corregidor auxiliar of the hamlet Molino T'ikanoma,[13] who travels frequently outside the province as part of union activities.

Thus, the sudden appearance of the SLI in Sacaca in 1995 was a dramatic and singular event for people in the province. Besides its legal functions, the Center also allowed women and their children, in effect, to live there permanently; part of the Center's funding included a subsidy for food services, which were contracted out to an adjacent family, which cooked for the women living at the Center. There were no time limits imposed on women and their children; the only restriction was that there had to be beds for every adult. The maximum capacity was seventy women, and the Center quickly filled up with women and their children. It turned out that some of the women who were able to stay the longest were the ones who had arrived the earliest; many stayed for more than six months while their cases were processed in the local court. But many women stayed at the Center even without actively pursuing their cases; while the legal aspects of their situations were often misunderstood, the fact that they could live in Sacaca away from abusive husbands or boyfriends meant that the Center's role became more complex than first envisioned.

Fortunately, the Center maintained accurate records of its activities, and the information contained in the SLI archive lends substance to claims about the dramatic impact on provincial legal consciousness of the Center's support for human rights. Although the Center served broad functions, its main purpose was to provide legal protection to women and children, and to this end Montesinos began presenting cases processed through the Center in front of the local judge very soon after the Center opened in 1995. Records from the *juzgado de instrucción* show that in 1995 the court opened forty-five new criminal cases, which represented almost a 50 percent increase over the average total number of criminal cases in the previous ten years.[14] Almost all of the nineteen additional cases above the previous ten-year average of twenty-six were cases filed by Montesinos on behalf of women who had come to the SLI.

The following tables give detailed information on the SLI's activities for the years 1995 through 1997.

Table 3.1 New Arrivals to SLI, 1995

No. of new arrivals	1st Quarter	2nd Quarter	3rd Quarter	4th Quarter	Total
Total	78	77	151	192	498
Women	59	68	104	169	400
Men	17	4	44	21	86
Children	2	5	3	2	12

Table 3.2 Types of Cases Processed by SLI, 1995

Type of case (W, M, C)*	1st Quarter	2nd Quarter	3rd Quarter	4th Quarter	Total
Physical aggression	30W	34W	52W	42W	158
Psychological aggression	42W, 3M, 2C	33W, 5M, 4C	60W	81W	230
Sexual aggression	4W	5W	6W	10W	25

* W = women, M = men, C = children; when no figure is given for a category here that means that none was given for it.

Table 3.3 New Arrivals to SLI, 1996

No. of new arrivals	1st Quarter	2nd Quarter	3rd Quarter	4th Quarter	Total
Total	389	201	159	111	860
Women	235	147	133	79	594
Men	154	48	21	27	250
Children	9	6	5	5	25

Table 3.4 Types of Cases Processed by SLI, 1996

Type of case (W, M, C)	1st Quarter	2nd Quarter	3rd Quarter	4th Quarter	Total
Physical aggression	69W, 1M	63W	38W	29W	200
Psychological aggression	314W	147W	109W	32W	602
Sexual aggression	6W	13W	7W	9W	35

Table 3.5 New Arrivals to SLI, 1997*

No. of new arrivals	1st Quarter	2nd Quarter	Total
Total	119	89	208
Women	58	48	106
Men	49	31	80
Children	12	10	22

* Statistics were not compiled by the SLI for the last two quarters of 1997 for unknown reasons. The statistics given here are for the year 1997 through June.

Table 3.6 Types of Cases Processed by SLI, 1997

Type of case (W, M, C)	1st Quarter	2nd Quarter	Total
Physical aggression	28W	31W, 8M	67
Psychological aggression	20W	13W	33
Sexual aggression	7W	5W	12

These tables show the remarkable impact of the SLI and the new ideas of legal identity it fostered within the province's legal universe. In its first year 400 women arrived at the Center. Out of this group, the types of violence complained of were roughly divided equally between physical and psychological aggression, with twenty-five cases of sexual aggression being distinguished from the general category of physical aggression. By its second year the SLI's impact was even more striking: Almost 600 women arrived at the Center in 1996. And as we can see, the number of cases of psychological aggression were three times the number of cases of physical aggression. In 1996 the number of cases of sexual aggression rose to thirty-five from twenty-five. But by the first two quarters of 1997, the numbers fell steeply: There are only 106 women as new arrivals, compared with 382 in 1996; and, most strikingly, there are only twenty cases of psychological aggression reported by women in the first quarter of 1997, compared with 314 in the first quarter of 1996.[15] The cases of reported sexual aggression, however, remain more or less constant between the first quarters of 1997 and 1996 despite the large drop in overall numbers between the two years.

Another intriguing feature of this data is the number of men who arrived at the Center between 1995 and 1997. In its first year of existence, eighty-six men arrived at the Center, compared with 400 women. In 1996 the number of men arriving to the Center tripled to 250, while the number of women also grew, but much more modestly, to 594 in 1996 from 400 in 1995. But by the first two quarters of 1997, the numbers of men and women arriving at the Center had become relatively equal compared with the previous two years. Yet despite the large numbers of men *arriving* at the Center, the number of cases processed with the three categories of aggression used by the Center was negligible. In 1995 there were eight cases processed involving men as victims of psychological aggression, in 1996 there was one case processed involving a male victim of physical aggression, and by 1997 there were eight cases processed involving men as victims of physical aggression.

Finally, the small number of cases involving children as victims of any of the three types of aggression should be noted. This is due to the focus of outreach efforts by SLI personnel. Although the protection of children is made an explicit part of the various pieces of legislation which led to the creation of the SLIs, the rights of children are subordinated to the rights of their mothers, as women. During outreach efforts in 1995, SLI personnel emphasized that the Center existed to protect the rights of women; children's rights were not distinguished from the rights of women, but were merely included as a subset of them. The idea was that if women came to the Center because of abusive domestic environments they would necessarily bring their smaller children, who could then receive legal protection and medical treatment if necessary. But it was felt that since the context in which the Center was created was so new and unsettling for many people in the province, that it was better to keep the Center's stated foci simple and direct without dividing families unnecessarily.[16]

But let us return to the first two observations about the SLI data: the dramatic rise and then fall in numbers of new arrivals, and the numbers of men arriving to the Center. The SLI had great success in spreading word of its existence and purposes during its first year of operation.[17] Yet despite the numbers of women—and men— coming to the Center, there were problems from the outset that contributed to the sharp decline—and thus influence—of the Center by 1997. First, although the Center continued processing cases in 1995 well after the Center itself had become full and not able to lodge more people, many of the women who could not use the Service as a refuge as well as a legal resource became discouraged. The Center was promoted as a place where women could find legal protection against physical and psychological abuse, but the sheer volume of new cases in 1995 simply overwhelmed the Center's staff of two.

The result was that in the following year fewer women decided to make the difficult decision to go to Sacaca and make their grievances public. The Center's inability to manage the number of new cases is demonstrated also through the records of the local court. Although there was a relatively dramatic spike in the number of new cases opened in the court in 1995 because of the Center's activities, the absolute number— 45—shows that only a very small percentage of the cases that merited legal intervention were actually prosecuted. And by 1996 and 1997, the number of new criminal cases in the court had returned to the prior ten-year average of twenty-six, indicating that the Center's impact in terms of criminal prosecutions was not significant.

The inability to meet the demands of new arrivals by prosecuting cases in the court is not only related to the lack of resources at the Center; the court itself served as an obstacle. The court had a staff of three and would not have been able to handle the numbers of cases that were coming through the Center during 1995 and 1996. But this was not the primary problem with the court; on most days the court did not have any activities scheduled and it could, if it desired, have increased its workload by at least 50 percent so that it would have been able to open approximately 300 new cases a month.[18] Private correspondence from the SLI archive indicates that the judge doubted the validity of both the mission of the Center and the merit of most of the individual cases brought to him by Montesinos. In some documents—mostly internal SLI memoranda—Montesinos complains bitterly about the resistance by the court to the Center's activities, and the fact that what he saw as overt hostility on the part of the judge meant that the Center was effectively prevented from functioning as it was designed. Montesinos also made formal complaints to the mayor during 1996 and 1997; although the mayor was particularly sympathetic to Montesinos's complaints, nothing was formally done and the Center was thus blocked from carrying out its mission.

The other striking feature of the information on the SLI shown through these tables are the numbers of men who arrived at the Center, relative to women, beginning in 1996. In 1995 the percentage of men arriving to the Center as a function of the total was 17 percent; in 1996 the figure was almost 30 percent; and, by the end of the second quarter of 1997, the figure had risen to almost 40 percent. In addition, the absolute number of men arriving rose significantly from 1995 to 1996—from 86 to 250, nearly a 200 percent increase—and by the end of the second quarter of 1997 the numbers were falling from the 1996 figures, although still well above the Center's first year. Already by the second quarter more men had arrived at the Center than in all of 1995. And even though the absolute numbers of men arriving at the Center were falling by mid-1997, the percentage of men to women continued to rise.

Yet of the 416 men for whom the Center recorded arrivals, only 17 managed to have formal cases opened on their behalf, which represents a mere 4 percent of the total. There are two explanations for this phenomenon. First, in their outreach efforts, the SLI was clear that the Center existed to provide a safe refuge and legal protection for women (and, secondarily, their children) who had been victims of physical or psychological violence. The legislation that created the SLI was intended specifically to address legal issues like sexual assault and domestic abuse, which were thought to be unique to women. As in the United States and other countries, women's rights activists in Bolivia did not deny the possibility that men were sometimes victims of domestic abuse;[19] rather, the view that the SLI legislation represents is that such incidents are comparatively rare and as such, not deserving of formal legal protection. But despite the fact that the SLI's mission was clearly articulated to hamlet authorities (during workshops in both Sacaca and the hamlets), who then passed on the information to people in their respective hamlets, something was lost in the process of transmission.

Because the opening of the Center was such an unprecedented event, many men in the hamlets assumed that they would benefit from its existence in some way, despite the official explanations that the Center existed to protect women's rights. To many men, the Center was a new and powerful legal institution in the province and one that would conceivably provide one more outlet for strategic legal maneuvering. Both men and women in the hamlets are sophisticated legal actors, particularly in the way the power of Sacaca's legal institutions is interposed between local hamlet authorities and competing local interests. Obviously the appearance of a new legal institution like the SLI—backed, as it was, by the combined power of the Church in

Sacaca, influential NGOs working in the region, and the national (not provincial) government—was certain to produce a degree of both excitement and disruption among men in the hamlets, and the large numbers of men who did arrive at the Center despite its stated intentions are proof of this.

But even though large numbers of men came to Sacaca and were officially recorded as new arrivals, new cases were not opened on their behalf; this is the second explanation for the discrepancy in the SLI statistics. When men arrived at the Center they soon learned, if they had not already known or were unclear, that it was not a general purpose legal institution like the local court. Moreover, in addition to not being a legal institution that could serve their interests, men who came to the Center discovered very quickly that their extended presence was not welcome. Because there were almost 100 women at the Center at any one time—most of whom were victims of domestic violence—the Center's staff was intent on preventing men from trying to seek revenge on spouses or partners who had left their hamlets and publicly denounced them. Despite the fact that new cases were not opened for the overwhelming majority of men who arrived at the Center, we are fortunate that the Center recorded their arrivals because such records are also—apart from the light they shed on the Center's activities—more strong evidence of the impact of transnational legal movements on legal consciousness in the province.

As I have mentioned above, the SLI was closed in August 1998. The house that the Center rented was returned to its owner, the offices were closed to women, and the records of the Center were archived. The simple explanation for why the Center closed is that after the Church in Sacaca relinquished its control over the SLI in Sacaca in April of 1998, it also ended its financial support. When the management of—and financial responsibility for—the Center passed to the mayor in 1998, much had changed, and it was only matter of time before the Center would run out of both money and governmental support. The Center was opened in 1995 during the Sánchez de Lozada administration, when the impetus for social reform was greatest; this was also the time when most of the progressive social legislation that authorized the SLI, and other reform institutions like Popular Participation, was passed. But in 1998 the political climate throughout Bolivia changed when a rehabilitated Hugo Bánzer Suarez swept into power;[20] as part of this change, the movement for social reform within the national government came to a virtual standstill, although bureaucratic entrenchment meant that agencies like SAG continued their existing work. New projects, however, were frozen. This meant that the SLI in Sacaca could not hope to obtain financial support from the departmental or national representatives of SAG.

At the local level, the new mayor in Sacaca—who entered office in 1998—was not nearly as sympathetic to the Center's mission as the previous mayor,[21] someone who was interested in social justice issues and who had supported the Center if not financially, then at least morally. And although the new mayor had doubts as to the validity of the Center's mission, his objections to it were more practical and less ideological than the objections of Sacaca's judge. During the same time the mayor was asked to take over the management of the Center, Popular Participation funds were just starting to flow into the town's coffers. The mayor had to make decisions about what to do with the money, and the way that Popular Participation monies were spent during 1998—all on infrastructure projects in Sacaca itself—indicates that he was very hesitant about distributing the funds in such a way that neither he as an elected official, nor the town, would clearly benefit.

But despite this failure, the pressing need for something like the SLI in Sacaca is evident, at least to those involved with the first project. Montesinos continued to receive at his house women who reported ongoing domestic violence as of August 1999, but Montesinos could not provide a refuge for them. He did continue to file claims on their behalf, but given that the judge remained unsympathetic to domestic violence cases, the number of cases filed by Montesinos after the Center closed remained close to what it had been before 1995. Nevertheless, Montesinos was insistent on finding new sources of nongovernmental financial support for a new Servicio Legal Integral for Sacaca; he even envisioned a series of SLIs throughout Potosí Department—Sacaca's had been the only one—with himself as regional director and adviser.

B. Lucio Montesinos as Moral Philosopher

The importance of Lucio Montesinos to Alonso de Ibañez's legal universe is difficult to overstate. He has had many roles as a local legal intellectual: civil registrar, town legal counsel, practicing lawyer, former director of the SLI, adviser to the Church on legal affairs, and former Agrarian Judge. But in the last ten years, his work with the SLI—especially the training in human rights and family law that accompanied it—has profoundly transformed his legal vision and his personal jurisprudence. Whereas before the 1990s Montesinos had always felt a certain amount of unease about the violence of social life in the province, he could not articulate his unease in concrete terms or within an explicit system of values. Moreover, he was suspicious that the frequently heard complaints about "underdevelopment" and "backwardness" in the province made by NGO workers and Sacaca townspeople only masked an underlying racism that he rejected.

But all this changed when he encountered Western human rights discourse for the first time. Here, finally, was a coherent system of legal values that seemed to express his own unarticulated intuitions about the inherent rights of individuals, especially those in society, like women and children, who had been more disadvantaged than others. Human rights law also appeared to Montesinos to be a politically neutral doctrine, conceived in the gentle and civilized courts and seminar rooms of Western Europe, far from the politically tainted legal universe in Bolivia. The source of human rights law also gave it a certain amount of prestige and influence analogous to the NGOs. But more than anything else, Montesinos's embrace of human rights doctrines allowed him, even if only intellectually or in the abstract, to transcend his humble surroundings and for a brief time join with the noble legal minds of other countries and other times. Although he viewed himself as firmly located at the ragged edge of human existence, far from the centers of legal power and movement, here he was, a man learned in the law, a *jurisconsulto*, someone who had come to the law late in life out of a love and respect for it against all popular opinion, a law which, as taught in Bolivia as in other civil law countries, is connected in an unbroken chain through the many centuries to the glorious perfection of Roman law. For Montesinos, his advocacy of human rights law was a way to resolve this contradiction.

From about 1995 on, Montesinos quickly incorporated his new jurisprudential vision into his work at all levels, slowly at first, as he attended seminars in La Paz on human rights law, but as his work at the SLI and training progressed, he adopted a more zealous approach. By the time I had arrived in Sacaca in 1998, Montesinos had refined his understanding of the meaning of human rights doctrine and was actively trying to single-handedly transform the legal consciousness of everyone in the province. His views were now quite clear: Human rights law—as expressed in national versions of international doctrines—defined a clear set of legal and moral imperatives that should govern the way people treated each other; that it was his job to make these imperatives known to people, both in Sacaca and in the hamlets; and that he was responsible for ensuring that once these imperatives were known and understood by people, they would be respected and followed on pain of legal sanction.

Because of his wide-ranging presence as a legal intellectual in the province, Montesinos had many opportunities to realize his complex jurisprudential visions in practice. Through many conversations with him, it was clear that he also saw himself as a moral philosopher much on the classical model. He felt that he possessed what amounted to a secret and superior knowledge and that it was his function to spread its benefits to those who were ignorant of it. The doctrine of human rights had become for Montesinos like a new religion to be preached to people who were in darkness. But human rights doctrine was not like a religion for Montesinos in a very important sense: It appealed to reason, not passion; logic, not faith. Human rights law was for him a superior secular answer to what he understood to be a religious-moral crisis in the province, one that was reflected in the omnipresence of domestic violence, an aberration that showed that the Church in Sacaca had not served its function, even though it had stood watch over the region for four centuries.

Montesinos is frequently asked to officiate at civil weddings in Sacaca and in some of the closer hamlets. As the only titled lawyer and one of only two civil registrars in the province, he receives approximately ten requests per month to perform wedding duties; although the sponsor of the wedding—the *padrino*—is obligated to offer Montesinos a small amount of money for his services, he does not accept any compensation for his work. Montesinos has officiated at almost all the weddings in Sacaca from the mid-1990s to the present, and during my time there he was increasingly being asked to officiate at hamlet weddings, especially in hamlets to which he was bound through *compadrazgo,* or fictive kinship, relationships. These events allowed him to begin the process of introducing the doctrine of human rights—especially rights related to the *rollos de género* ("gender roles"), his particular concern—at a very important moment in people's lives and with maximum dramatic impact.

I was present at several weddings in Sacaca and one in a hamlet—Kamacachi—and Montesinos's routine did not vary. Almost everyone in the province has a civil wedding because the costs associated with a ceremony in the Church and the party that follows it are prohibitive for most people. As in other parts of the Andes, the couple to be married asks another couple to sponsor the wedding and these godparents, not the couple's parents, will perform the most important functions during the civil ceremony. There are no formal obligations regarding the length or content of a civil service; the only requirement is that the godparents sign the wedding certificate as witnesses and that the signing be supervised by an authorized official such as a titled lawyer, civil registrar, or judge. But despite the lack of official requirements involved with conducting a civil wedding in Bolivia, and the guests' preference that the formalities be kept to the absolute minimum so that the *ch'allas,* or ritual toasts, and dancing can begin, Montesinos has developed a wedding procedure that rivals a church mass in its seriousness of purpose and length.

As soon as he arrives at the wedding site—which is in most cases the family *sala de recepción,* a room with a dirt floor and low benches along all four walls—he proceeds immediately to the wedding table and spreads out his instruments: the wedding certificate, a copy of the Bolivian family code, a copy of the Bolivian civil code, and a copy of Law 1674, which he will draw on for much of the human rights language that will figure prominently in the ceremony. Montesinos is always eager to begin the ceremony as soon as possible because he knows that for many people it will last longer than expected and will feature the introduction of ideas and concepts that they will not have anticipated. As soon as the couple is standing in front of his table—flanked by their godparents—Montesinos begins reading from the Bolivian law codes' sections on marriage, which merely outline the procedural requirements for a legitimate service. But that is soon finished and then he begins the next stage by intoning solemnly "existen en este mundo derechos humanos" ("there exist in this world human rights"). He then describes these human rights as "universal laws" that define responsibilities that the newly married couple has to themselves, their future children, and to society.

After spending about thirty minutes providing this introduction to human rights doctrine, Montesinos then moves on to a subset of these rights, and his particular interest, the *rollos de género.* He lectures the couple that the rules governing relationships among men and women in the province have changed significantly since the time of the *antepasados,* or ancestors, and that now the man's duties toward his wife included helping her cook for the family, helping with childcare—including carrying children on the trail, a burden women complain about frequently—and devoting equal time to cleaning around the house. But most of all, a man can no longer use violence against his wife or children for any reason. Montesinos explains that both international human rights law and Bolivian law now protect women from abuse; he even goes into detail about specific acts prohibited under Bolivian law and the amounts of prison time associated with each. Montesinos also devotes some time to discussing the fact that women can also be perpetrators of violence in relationships, but he typically emphasizes that women usually resort to psychological and not physical violence.

The preparatory lecture on human rights in general, and then the specific discussion of gender roles, together last about one and a half hours. Because of the seriousness of the occasion and the reputation of Montesinos, the wedding party usually does not do anything to interrupt or otherwise show disrespect for him. But as the minutes roll by and the discussion about men's duties continues seemingly without end, the people become noticeably restless and uncomfortable and it is difficult to say how much of Montesinos's discussion is actually heard and understood by people present. The wedding couple, however, always seems to listen intently, if only because they are directly in front of Montesinos and his words are directed to them. Finally, after about two hours—which is about an hour and fifty minutes longer than the average Bolivian civil wedding ceremony—the wedding couple is released from Montesinos's grip, they and their godparents sign the certificate, and the party begins.

Since Montesinos began this human rights-infused civil wedding procedure in 1995, he has presided over approximately one hundred weddings. Because he will continue to use this approach as long as he is asked to officiate at weddings in the province, and because he remains the only person regularly asked to do so, it seems likely that an entire generation of young people will receive his instructions on human rights at this crucial moment in their lives. It is difficult to assess the impact of this moral philosophizing on young people, but right before I left Sacaca in August 1999 a mural that was created by an organized group of some seniors from the local high school appeared just off Sacaca's plaza. The mural said: "Organizados por nuestros derechos" ("We are organized for our rights"); "derechos" in this context clearly meant "derechos humanos."

Apart from civil weddings, which are the most obvious and dramatic settings in which Montesinos pursues his passion for teaching others about human rights, there are also other areas in which he is able to insert his moral and legal vision into his work. During consultations in his office as a private lawyer, Montesinos frequently refers to general principles of human rights—and especially the rights of women and children—when advising clients and planning legal strategies for sessions at the local court. Although he does not have the personal resources to restart the SLI, he does continue to receive women in his office who are victims of the same types of crimes handled by the Center: physical and psychological abuse, sexual assault, spousal abandonment, and breach of promise to marry in exchange for sexual relations. Montesinos accepts as many of these cases as he can given his schedule and other duties.

The last major area of his work in which Montesinos is able to incorporate his moral and legal vision is during court sessions. Because of the contentious nature of the relationship between Montesinos and the current judge—which is linked in large part to the judge's resistance to the SLI's activities—court sessions in which Montesinos represents a party can devolve into a tense battle of wills, particularly when the case involves sexual assault, physical aggression, or spousal abandonment, types of cases to which Montesinos devotes most of his time and energy. What frequently happens is that it becomes apparent soon after a court session begins that the judge is not favorably disposed to the claims of Montesinos's client, who is usually a plaintiff.

But before allowing the judge to dismiss the claim and end the session, something Montesinos is prepared for, he asks politely if he may have a chance to make a formal statement. Because the request is made so calmly and formally despite the obvious tension between them, the judge usually grants Montesinos's request. This gives Montesinos the opportunity to deliver a version of his human rights/gender roles lecture that he delivers at weddings; only here the lecture is not given to a couple as they look to the future, but rather to a man (or boy) who has already violated the moral order as Montesinos understands it. Despite knowing that his client will receive no legal redress, when he is allowed to address the court in this situation Montesinos delivers a fiery lecture and rebuke until the judge finally ends the session by ringing a small bell. I have seen many younger men and boys leave the room weeping, with Montesinos closely following and still admonishing them for violating what he calls "el derecho universal," the universal law.

Implications for Methodology

The types of legal movements described above clearly call for a much broader set of research methods than simply localized participant-observation and interviewing, because the local–regional–global articulations are so apparent and important. Given the changes associated with globalization—especially the ongoing integration of capital, markets, technology, and information across national borders—the legal ethnographer can no longer assume that a study focused on either local legal institutions or actors (or both) will be sufficient. Rather, legality must be conceptualized as both more fluid and unstable, a shifting set of normative practices and ideas that form a network that is mostly invisible.

One way to understand the social forces driving the local–regional–global articulations of law is through Appadurai's concept of "cascading" (1995), in which the different levels of a social network can become sudden channels for the rapid and unpredictable movement of ideas and practices. Appadurai uses "cascading" to describe the way in which beliefs about ethnic nationalism move between the transnational and the most local of levels, sometimes with violent and unintended consequences. The movement of legal ideas in an era of global interconnections can be understood in much the same way. There are really two major methodological problems that flow from this reconceptualization: first, the challenge of actually following these movements, given the obvious practical difficulties; and second, the difficulty in observing the *effects* of these globalized legal movements, in light of the essentially unpredictable movement of legal ideas and practices through what is at best a translucent social network. During the course of the research that allowed me to track the changes described above, I experimented with two techniques as a solution to this two-part challenge.

A. Follow the Ideas

If it is true that in order to uncover the primary motivations behind political behavior one should follow the money, then it is also true that if a legal ethnographer wants to understand the complexity of legality in an era of globalization, he or she should follow the ideas, and by ideas I mean both formal legal theory and notions of what constitutes proper legal behavior. A very good way to do this is to adopt some version of the multisited approach to ethnography urged by Marcus (1995, 1998) and others. In this approach the legal ethnographer assumes that "the field" does not denote primarily a bounded location in space—such as a village, courtroom, or legal district—but rather a set of relationships which are linked by common interests.[22] This means that the ethnographer must follow the legal ideas where they lead, and they often lead to unexpected places.

In tracking the movement of legal ideas from, to, and within rural Bolivia, I had to likewise move constantly. Participant-observation and interviewing still formed the foundation for this multisited research, but I traced the movement of human rights doctrine by interviewing NGO representatives not only in Alonso de Ibañez, but throughout Bolivia, including at their headquarters in the nation's capital. Only financial considerations prevented me from following the legal ideas back to Europe or the United States. I followed union workers, like the corregidor auxiliar of Molino

T'ikanoma (see above, note 13), as he made his way from union meeting to union meeting and to larger regional centers, where he agitated for workers' rights and soaked up more rights doctrine in the process.

The legal ideas that were important for my research made their way into documents, and so I followed the paper trail as it took me to Sucre, the nation's legal capital, and Potosí, the relevant departmental capital. In making legal ideas the central objects of research, as opposed to institutions or actors, my focus obviously changed from the traditional legal ethnographic project. Further, the research process became more unpredictable, because I did not know in advance where the movement of legal ideas would lead or even if they would continue to be dynamic. But the reward was that the project gained a coherence that did not seem artificial or contrived. It did in fact appear to me that the ideas themselves were motivating disparate actors in ways that I could trace and understand using these multisited techniques. Conversely, it would have been difficult to track the movement of, and understand, these ideas by treating the legal universe of Alonso de Ibañez as closed, "autopoeitic," or structurally static, or the "field site" as circumscribed by the boundaries of the province.

B. Legal-Intellectual Biography

The second technique that I used to track the impact of globalization on legality in Alonso de Ibañez was to focus on what can be called "legal-intellectual biography." Although there are many legal actors in the province, it became clear to me from early on that certain individuals in the province were serving as lightening rods for the movement of legal ideas. Given the fact that the province itself was at the receiving end of so many international development efforts, it was perhaps inevitable that a few people would have realized from the beginning that new (in this case legal) ideas would from then on be transforming local understandings in profound ways. Lucio Montesinos was one such person. Because I was also attempting to follow the important legal ideas, I realized that I was confronted with something of an epistemological tension: On the one hand, I was emphasizing the way legal ideas moved in dynamic and unpredictable ways and how these processes were central to producing knowledge about legality in Alonso de Ibañez; but on the other hand, by focusing on legal intellectuals like Montesinos, I was moving away from the legal ideas themselves and (appearing to) return to a traditional legal ethnographic focus on legal actors. At the time this tension caused me much frustration, but I see now that it is endemic to the current problem of using ethnographic methods to study legality through its local–regional–global articulations.

Because of developments in technology and communication, legal ideas like rights doctrines could be announced in Europe or the United States (for example ILO Convention 169), transmitted to La Paz, then introduced to someone like Lucio Montesinos, who then returns to Alonso de Ibañez and introduces such ideas to the thousands of people in the province. Within a short time the legal ideas are transformed locally, and representatives of NGOs, or Montesinos himself, then serve as channels for the flow of the new legal ideas back to the sources. For example, the notion of "indigenous rights," which NGOs attempted to incorporate into development efforts in rural Bolivia, underwent significant changes as such ideas were introduced locally and then were altered to meet local expectations. These modified understandings of indigenous rights were soon being debated at NGO headquarters in Brussels, or Amsterdam, or New York City, and NGO policies would be redesigned accordingly.[23] This process unfolds rapidly, at times in less than one month, and legal intellectuals like Montesinos play disproportionately significant roles.

For this reason, the legal ethnographer interested in tracking the rapid movement of legal ideas as they make their through the various points in a global–regional–local network gains by focusing on the work of "superempowered individuals" (Friedman 1999) like Montesinos. It is not my contention that the forces of globalization require legal ethnographers to adopt a "great person" approach to fieldwork. But as my research has shown, certain individuals, especially in areas like rural Bolivia which are targeted for the introduction of transnational legal ideas like rights doctrines, serve as the primary engines driving the process at the local level, and the legal ethnographer should therefore make their life and work central objects of study.

Conclusion

In this chapter I explored the extent to which globalization is changing the contexts in which many legal ethnographers conduct research. Through an examination of the impact of the arrival of Western human rights discourse to rural Bolivia during the last ten years, and the role played in this process by certain "superempowered" legal intellectuals, I was able to show why I was forced to supplement conventional ethnographic methods with two techniques not normally included in the traditional canon: the use of multiple research sites, and a focus on legal-intellectual biography. Although legal ethnographers will need to continue to find new methods to add to participant-observation and interviewing in single sites in response to the challenges of conducting legal ethnographic research in an era of globalization, both of the techniques I adopted in order to track the complex movement of legal ideas proved effective and could be used in other similar research environments.

Notes

This chapter is the result of fifteen months of research in Bolivia completed in 1999, which was made possible by the generous support of the National Science Foundation, Law and Social Science Program (SBR# 9807836), the Organization of American States (F57035), the HEA Title VI Foreign Language and Area Studies Program, and the David L. Boren Graduate Fellowship Program. I would also like to thank my former colleagues in the Department of Anthropology, University of Wisconsin-Madison, and in the Institute for Legal Studies, University of Wisconsin Law School, for many helpful suggestions regarding the ideas in this chapter.

1. Other segments of Bolivian society—especially the Army and the financial elite—remained hostile to the upsurge in indigenous mobilization because of the perceived threat to either social stability (for the Army) or economic stability (for the moneyed classes).
2. The countries that had ratified Convention 169 as of 2000 were the following: Mexico (1990), Norway (1990), Bolivia (1991), Colombia (1991), Costa Rica (1993), Paraguay (1994), Peru (1994), Honduras (1995), Denmark (1996), Guatemala (1996), Ecuador (1998), Fiji (1998), Netherlands (1998), and Argentina (2000) (www.ilo.ch).

3. Convention 169, officially called the "Indigenous and Tribal Peoples Convention, 1989," has forty-four articles and was meant to either replace or update many provisions of the "Indigenous and Tribal Populations Convention, 1957." Convention 169 defines indigenous "peoples" (which replaces "populations") in Article 1, Section 1(b) as "peoples in independent countries who are regarded as indigenous on account of their descent from the populations which inhabited the country, or a geographical region to which the country belongs, at the time of conquest or colonization or the establishment of present state boundaries and who, irrespective of their legal status, retain some or all of their own social, economic, cultural and political institutions." Among the forty-four articles, two are particularly relevant for my purposes here. Article 8 reads in full:

1. In applying national laws and regulations to the peoples concerned, due regard shall be had to their customs or customary laws.
2. These peoples shall have the right to retain their own customs and institutions, where these are not incompatible with fundamental rights defined by the national legal system and with internationally recognized human rights. Procedures shall be established, whenever necessary, to resolve conflicts which may arise in the application of this principle.
3. The application of paragraphs 1 and 2 of this Article shall not prevent members of these peoples from exercising the rights granted to all citizens and from assuming the corresponding duties.

And Article 9 reads in full:

1. To the extent compatible with the national legal system and internationally recognized human rights, the methods customarily practiced by the peoples concerned for dealing with offences committed by their members shall be respected.
2. The customs of these peoples in regard to penal matters shall be taken into consideration by the authorities and courts dealing with such cases.

4. For how NGOs working in the Andes have attempted to utilize "saberes andinos," see especially Cueto 1995. The conservation and development literature on developing countries and the Andes is extensive, but see in particular two very good studies from a geographer's point of view: Zimmerer and Young (eds. 1998), and Zimmerer 1996. The definitive work critical of "development" as an idea and ideologically-charged practice is still Escobar 1995.
5. Some of the NGOs that were active in rural Bolivia during 1990s that adopted this approach were the following: Mosoj Causay (Belgium, potable water), PCI (United States, "work for food"), and PROINPA (Holland, potato projects).
6. During 1998–1999 I was present during many of these initial "context-building" meetings. Most of the official descriptions by NGO leaders were simply versions of relevant portions of international charters, for example ILO 169, Article 7, Section 1, which reads:

1. The peoples concerned shall have the right to decide their own priorities for the process of development as it affects their lives, beliefs, institutions and spiritual well-being and the lands they occupy or otherwise use, and to exercise control, to the extent possible, over their own economic, social and cultural development. In addition, they shall participate in the formulation, implementation and evaluation of plans and programmes for national and regional development which may affect them directly.

7. An "ayllu" is an ethnic political and legal entity that has been the basic unit of Andean social organization from pre-hispanic times. Particularly in the north of Potosí, ayllus retain many prehispanic features, including "an internal organization based on dual and vertically-organized segments, communal distribution of resources, and a 'vertical' land tenure system which includes the use of non-contiguous *puna* (highland) and valley lands" (Rivera Cusicanqui 1991; see also Platt 1982). The internal organization of ayllus in the north of Potosí can be conceptualized as a set of inlaid boxes, with each territorial and kinship unit

part of an ever larger set of ethnic units, which culminate in one grand unit, itself divided into two moieties, which relate to each other as complementary opposites (Platt 1982: 5).

8. "Hamlet" is the best word in English to describe the aggregations of families who live outside Sacaca in the province. The words "town" or "village" convey a sense of size and structure that is inappropriate as applied to these aggregations. The hamlet dwellers themselves use words in either Quechua or Aymara to describe where they live; in Quechua, the word "llajta" is used, preceded by the name of the hamlet, for example "Jankarachi llajta." But llajta is best translated as "place," which is not definite enough for wider application. The Spanish words used in Alonso de Ibañez for the hamlets are either "ranchu" (a Quechuazation of "rancho") or "estancia" ("farm or cattle ranch"); "comunidad" ("community") is also sometimes used. Because these Spanish words are used in legal documents, they have been widely adopted by the people in the hamlets themselves, to the point where they have replaced "llajta" with either "ranchu" or "estancia" when discussing their hamlets among themselves.

9. Although both "ranchu" and "estancia" are used interchangeably by the authorities in Sacaca and at the regional and national levels to refer to the ayllu hamlets in Alonso de Ibañez, the terms are not synonymous to the hamlet dwellers themselves, a fact that seems to have been overlooked by both Bolivian census workers and researchers (both Bolivian and foreign). To the *runa* ("the people" in Quechua, the term used by people in many parts of the Quechua-speaking highlands to refer to themselves in their own language, which they call *runa simi*, "language of the people"), both ranchu and estancia can refer to their hamlets according to common usage (see note 8). But the word ranchu is reserved for the bigger of the hamlets, usually one of the major cantonal centers. Estancia is used for everything else. The confusion lies in the fact that strict guidelines are not used when calling one hamlet a ranchu or an estancia; one just "knows" which hamlets are ranchus and which are estancias, and this intuitive knowledge can only be accessed by asking people in as many hamlets as possible. Although I did research in forty out of the approximately 200 hamlets, I am still unable to list a definite set of criteria in this regard. Despite this, it is possible to say that most of the hamlets that are called ranchus have thirty or more families in them (the unit of measurement for people when describing the size of their hamlets).

10. The clearest example of this involves the most controversial part of the reform legislation: the 1996 Ley de Tierras INRA (Instituto Nacional de la Reforma Agraria), which was intended to replace much of the important aspects of the agrarian reform law that developed after 1952. The neoliberal orientation of INRA has been subjected to trenchant criticism in one important study (Antezana 1999); in another, it receives a better review, yet still within the context of much skepticism as to both the motivations behind it—the desire to make land available for large centralized companies—and its effects—less participation of peasants in decisions over land (Solón 1997).

11. In 1998 Hugo Bánzer Suarez swept into power largely on the vote of the Santa Cruz region, where he had often sought refuge during his many political intrigues dating from the early 1970s.

12. Law 1493 (17 September 1993), a law passed through Bolivia's executive branch, created the Ministry of Human Development. Article 71, No. 5 of Supreme Decree 23660 (12 October 1993) created the National Secretariat for Ethnic and Gender Issues. Articles 85, 86, and 87 of this same Supreme Decree created the Subsecretariat for Gender Issues responsible for all political matters related to women. The Ministry of Human Development, in Resolution 139/94 (21 September 1994), adopted the National Plan for the Eradication, Prevention, and Punishment of Violence Against Women. Article 1 of this Resolution created the system of Servicios Legales Integrales to carry out the Resolution's objectives. Law 1674 (1995), passed by the Bolivian Congress, outlined the nature and function of the SLIs and authorized their establishment. In practice, SLIs can

only be established after a formal application is made on behalf of a municipality with the assurance that supplemental funding will be provided.

13. A corregidor auxiliar is a "traditional" authority position found in some parts of rural Bolivia that dates to the colonial era. This particular corregidor auxiliar was famous in Alonso de Ibañez for his union activism and for the way he studied international human rights doctrine and proselytized with these ideas to other union members throughout the province, and indeed throughout the north of Potosí Department. For more on him, see Goodale 2001.

14. The numbers of new criminal cases opened in the local court, the *juzgado de instrucción*, between 1985 and 1994 are as follows: 1985 (38), 1986 (31), 1987 (29), 1988 (25), 1989 (23), 1990 (21), 1991 (30), 1992 (16), 1993 (23), 1994 (27).

15. Yet despite the drop in numbers, the total number of women who came to the Center between 1995 and the end of the 2nd quarter of 1997 as a percentage of the total number of women in the first section of the province is still significant. In fact, intake forms from the SLI archive show that most of the women were from the province's first section (there are two sections in the province), in which case the 1,100 women arrivals would represent fourteen percent of the female population of the first section based on figures in the 1992 National Census (INE 1992). But even if we use the female population of the entire province, the number of women who came to the Center complaining of domestic abuse represented nine percent of the total population, still a relatively high figure.

16. Having said this, it is important to note that the absolute number of new cases involving children did rise steadily from 1995 to 1997, although the percentage of new cases involving children was small compared to cases involving women and, somewhat ironically, men. In 1995 there were 10 cases involving children; in 1996 there were 25 cases; and in 1997 there were already 22 cases involving children by the end of the second quarter.

17. This was done through workshops in Sacaca, workshops in individual hamlets, particularly the important cantonal centers, and, necessarily, by word of mouth. During the workshops in both Sacaca and in the hamlets, local authorities were officially invited to participate and then, after being instructed as to the Center's intentions, asked to collaborate with the Center by enforcing Law 1674 and by encouraging women to come the Center as needed. The following workshops were held: Sacaca (June 1995), Sacaca (April 1996), Layupampa (second quarter 1997), Llapa-Llapa (second quarter 1997), Waraya (second quarter 1997), Sacaca (July 1997), Sacaca (September 1997), and Tarwachapi (November 1997).

18. This does not mean that there would have been twenty-five hearings a month, which would have been difficult; but the court could have opened twenty-five *new cases* a month without much difficulty.

19. Though not sexual assault of men by women, the possibility of which would be unthinkable in Bolivia.

20. He had previously ruled the country as a military dictator—then known as Colonel Banzer—between 1971 and 1978.

21. The new mayor did not support the Center even though he was a member of Sánchez de Lozada's MNR (Movimiento Nacionalista Revolucionario) party. Although Banzer's party is the ADN (Acción Democrática Nacionalista), there is often a considerable time lag before the new political party at the national level begins replacing functionaries at the provincial level.

22. In legal anthropology, Sally Falk Moore's semi-autonomous social fields (1973, 1978) certainly required a reconceptualization of "the field" much along the lines discussed in Marcus, and Gupta and Ferguson (1997a,b), and even Santos (1995).

23. I would like to thank the Belgian director of the NGO Mosoj Causay, which was active in Alonso de Ibañez during 1998–1999, for information regarding the type of process described here.

References

Archives consulted
Sacaca:
Juzgado de instrucción (JDI)
Servicio Legal Integral-Sacaca (SLI)
Corregidor titular de Canton Sacaca (cuaderno de actas), 1996–1999 (ACT)
Director provincial de policía de Sacaca, 1998–1999 (DPP)
Hamlets:
Cuaderno de actas, Molino T'ikanoma, minor Ayllu Jilawi Cuerpo/Mayor (CDAMT)

International laws consulted
International Labor Organization, Convention 169 (1989)

Bolivian laws consulted
Ley No. 1551 (1994), Participación Popular
Ley No. 1654 (1995), Descentralización Administrativa
Ley No. 1674 (1995), Servicios Legales Integrales
Codigo civil
Codigo de familia

Published sources
Antezana, Luis E.
    1999   Trampas y mentiras de la Ley INRA. La Paz, Bolivia: Editorial Juridica "Temis."
Appadurai, Arjun
    1996   Modernity at Large: Cultural Dimensions of Globalization. Vol. 1, Public
    Worlds. Minneapolis: University of Minnesota Press.
Barrios de Chungara, Domitila
    1978   Let Me Speak! Testimony of Domitila, A Woman of the Bolivian Mines.
    New York: Monthly Review Press.
Bolivia
    1992   National Census. Potosi: INE.
Cueto, Marcus, ed.
    1994   Missionaries of Science: The Rockefeller Foundation and Latin America.
    Bloomington: Indiana University Press.
Cueto, Marcos, ed.
    1995   Saberes andinos: ciencia y tecnologia en Bolivia, Ecuador y Peru. Lima: Instituto
    de Estudios Peruanos.
Escobar, Arturo
    1995   Encountering Development: The Making and Unmaking of the Third World.
    Princeton, NJ: Princeton University Press.
Friedman, Thomas
    1999   The Lexus and the Olive Tree: Understanding Globalization. New York: Farrar,
    Straus, Giroux.
Goodale, Mark
    2001   A Complex Legal Universe in Motion: Rights, Obligations, and Rural-Legal
    Intellectuality in the Bolivian Andes. Doctoral dissertation, University of Wisconsin-
    Madison.
Gupta, Akhil and James Ferguson, ed.
    1997a   Anthropological Locations: Boundaries and Grounds of a Field Science.
    Berkeley: University of California Press.
Gupta, Akhil and James Ferguson, ed.
    1997b   Culture, Power, Place: Explorations in Critical Anthropology. Durham, NC:
    Duke University Press.

Marcus, George
    1995   Ethnography in/of the World System: The Emergence of Multi-Sited Ethnography. Annual Review of Anthropology 24: 95–117.
    1998   Ethnography through Thick and Thin. Princeton, NJ: Princeton University Press.
Moore, Sally Falk
    1973   Law and Social Change: the Semi-Autonomous Field as an Appropriate Subject of Study. Law and Society Review 7: 719.
    1978   Law as Process: An Anthropological Approach. London: Routledge and Keegan Paul.
Platt, Tristan
    1982   Estado boliviano y ayllu andino: tierra y tributo en el norte de Potosi. Lima, Peru: Instituto de Estudios Peruanos.
Rivera Cusicanqui, Silvia
    1991   Liberal Democracy and Ayllu Democracy in Bolivia: The Case of Northern Potosi. Journal of Development Studies 26(4): 97–121.
Santos, Boaventura de S.
    1995   Toward a New Common Sense: Law, Science and Politics in the Paradigmatic Transition. New York: Routledge.
Solón, Pablo
    1997   ¿Horizontes sin tierra? Analisis critico de la Ley Inra. La Paz, Bolivia: Cedoin.
Zimmerer, Karl
    1996   Changing Fortunes: Biodiversity and Peasant Livelihood in the Peruvian Andes. Berkeley: University of California Press.
Zimmerer, Karl and Kenneth Young, eds.
    1998   Nature's Geography: New Lessons for Conservation in Developing Countries. Madison: University of Wisconsin Press.

Chapter 4

Analyzing Witchcraft Beliefs

Jane F. Collier

In 1967–1968, when I was doing field research for my dissertation on "law" in the Tzotzil Maya community of Zinacantan, Chiapas, Mexico, I periodically took time off from collecting cases to explore a particular topic in depth. I spent a month focusing on marital problems, collecting accounts of all the divorces and reconciliations that had occurred within memory in one Zinacanteco hamlet of 600 people. The two months I spent analyzing witchcraft beliefs, however, were the most fruitful by far. Analyzing witchcraft helped me not only to solve puzzles that had arisen in my analysis of cases, but also gave me a way of fitting my growing understanding of Zinacanteco law into the theoretical shifts that were occurring in American anthropology in the 1970s and early 1980s.

Zinacantan is a rural municipality that in the 1960s had a population of approximately 9,000 Tzotzil-speaking Indians, and whose population has since grown to around 30,000 in the late 1990s. It is one of several indigenous municipalities surrounding the highland city of San Cristobal de Las Casas, which is populated primarily by "Ladinos" (the local term for Spanish speakers of mixed Indian and Spanish descent). The municipality of Zinacantan is quite large, covering approximately one hundred square miles of mountains and valleys. Most Zinacantecos live in scattered hamlets, in small houses that were surrounded by cornfields in the 1960s and that are today surrounded by fruit orchards. The municipal and ceremonial center of Zinacantan, known as Hteklum in Tzotzil, is in the northernmost high valley. It contains the two main churches (both Catholic) and the town hall. It is also surrounded by the most important sacred mountains where shamans pray to the ancestor gods.

In the 1960s, the population of Hteklum was very small. It had around 400 residents, many of whom were living there temporarily while serving in religious or civil offices. In the 1960s, Hteklum also had one street lined with small stores owned by Ladinos. Today, in the late 1990s, few Ladinos live in Hteklum. They have sold their stores to Zinacantecos. The population has more than tripled and today Hteklum has many paved streets. A new and imposing town hall has replaced the long, low building whose open porch served as the municipal court in the 1960s. The court is still staffed by elected Zinacanteco officials who serve three-year terms, but in the 1960s all the civil officials heard trouble cases, whereas today only those elected as judges, or *regidores,* staff the court. The municipal president, who chaired the court in the 1960s, now sits in on cases only when very important people or issues are involved. The court has also moved indoors, out of the view of casual passersby. When I attended court in the fall of 1997, it was held in one room of the spacious town hall. When I returned in the fall of 1998, the judges had moved to a new courthouse behind the town hall, which had just been built by the state government. But the judges I observed did not use the courthouse as its builders intended. They seldom heard cases in the large and imposing courtroom. Instead, they preferred to hold court in a small back office where they could speak with disputants face to face rather than having to look down on them from a raised dais (Collier 1999).

In the 1960s, I collected most of the data for my dissertation by interviewing paid informants at my home in the city of San Cristobal. I did sit in on a few cases at the town hall court in Hteklum. But because I was a young woman studying a court run by men, I found that sitting around the town hall waiting for cases to arrive was awkward both for me and for the town officials. My female presence disrupted their leisure pastime of sexual joking.[1] Moreover, I found it hard to understand the cases that I did observe. My command of Tzotzil was not up to following rapid and often angry interchanges. But if I found it hard to observe cases, I was able to conduct long interviews with Zinacanteco men who had been municipal officials or who were locally renowned as popular mediators. I was a member of the Harvard Chiapas Project, directed by Professor Evon Z. Vogt, and thus able to take advantage of all of the project's resources, which included the network of good relations that project members had built up with important Zinacantecos. The project had developed the policy of paying informants for formal interviews, based on the idea that it was only fair to compensate busy people who took time off from their money-earning occupations to answer questions. We did not pay people for the time we spent hanging out with them, however, although we did try to reciprocate people's hospitality with gifts of food and with transportation at a time when no Zinacantecos owned trucks or cars.

The first puzzle that confronted me when I began to interview informants about Zinacanteco law was that I could not get them to provide me with lists of crimes and punishments. They seemed unable to answer general questions about how offenses such as murder, theft, or assault were handled, although they had no difficulty recalling particular cases. When telling me about cases, however, they seemed less interested in talking about what individuals had done than in explaining to me the relationship between the parties. From their point of view, the relationship seemed to determine both the meaning of individuals' actions and the range of possible solutions to a conflict. This focus on relationships rather than actions puzzled me, because the books I had read on legal anthropology led me to believe that every community has "laws," in the sense of having some norms whose "neglect or infraction is regularly met, in threat or in fact, by the application of physical force by an individual or group possessing the socially recognized privilege of so acting" (Hoebel 1954: 28; see also Pospisil 1958). By analyzing the cases I collected from informants, I found that I could make up a list of offenses and likely punishments. But my analysis of cases also convinced me that Zinacanteco informants were right when they treated the relationship between the parties as more important than the actions of individuals for determining the outcome of a case.

The second and more interesting puzzle arose from my attempt to solve the first one. After I realized that Zinacanteco informants talked about cases more in terms of managing relationships than in terms of punishing wrongdoers, I started to use the methodology proposed by Paul Bohannan in his 1957 book, *Justice and Judgement among the Tiv,* which concerns a tribe in Nigeria. Bohannan wrote his book as an argument against Max Gluckman, whose analysis in *The Judicial Process among the Barotse of Northern Rhodesia* (1955) concluded that judges everywhere reason in much the same way because judges everywhere face a similar task. Bohannan argued instead that judicial processes are culturally specific: Judges reason differently according to how they understand their task. An ethnographer must therefore capture native understandings. Bohannan focused on analyzing the Tiv words that litigants and judges used to talk about what they were doing. Following Bohannan, I thus set out to isolate and analyze the Tzotzil words that Zinacantecos used to talk about the disputing process. I collected texts written in Tzotzil, and translated them with the help of informants and the exhaustive Tzotzil Dictionary compiled by Robert Laughlin (1975).

Once I started to analyze key words, I discovered that Zinacantecos used the same term—"hsa' k'op" (dispute seeker)—to refer both to the person who picked a fight and to the person who took a case to court to try to settle it. This puzzled me because my common sense suggested that the two roles are incompatible. In what kind of imagined world, I wondered, could someone who starts a fight be viewed as fundamentally similar to someone who tries to end one? I knew that other members of the Harvard Chiapas project (Vogt 1965, 1969) and ethnographers who worked in other highland Chiapas communities (Guiteras Holmes 1961; Hermitte 1964) had analyzed the "world view" of indigenous peoples as rooted in cultural beliefs about the soul. But I could not see any direct connection between soul beliefs and legal processes. After all, I had been taught to distinguish law from religion by the theoretical texts I read in legal anthropology. Those texts had all separated sacred from secular institutions, although often portraying them as complementary, mutually supporting ways of maintaining social order (Evans-Pritchard 1976). Hoebel, for example, assumed a distinction between "sins" and "crimes" as he argued that supernaturally imposed sanctions could complement the sanctions imposed by socially recognized authorities (1954).

The anthropological texts I read also tended to distinguish religion from magic and witchcraft (Malinowski 1948). Religion was portrayed as offering answers to such existential questions as the meaning of human life, whereas both magic and witchcraft were considered practical tools for influencing events. Some authors who distinguished religion from magic and witchcraft argued that the latter two, but not the former, would gradually disappear as humans developed better means for solving practical problems. Science would replace magic (Malinowski 1948), and law would replace witchcraft. Hoebel, for example, argued that "law is the natural enemy of sorcery"—that witchcraft flourishes only "where the overt mechanisms of law have not been worked out." He also thought that "magic," although long remaining "the handmaid of the law, mopping up where the broom of the law fails to sweep clean," would also fade away as the investigative and punitive powers of judges increased (Hoebel 1954: 274).

When I began to study Zinacanteco law, I too distinguished law from religion and witchcraft. As part of my general project of collecting cases, I collected witchcraft disputes, but I treated witchcraft as simply another kind of offense that Zinacantecos could commit. I analyzed the cases I collected to find out what kinds of people were commonly accused of witchcraft and by whom, how cases of witchcraft became public and reached the attention of secular authorities, and how such cases tended to be settled. I also treated witchcraft as another way of seeking justice, following Nader, who had included "witches" among the "remedy agents" available to aggrieved individuals in the Zapotec community she studied (Nader 1969). Like the legal anthropologists whose works I had been reading, I thought of witchcraft as an alternative to law, not as integral to it. My breakthrough into understanding how Zinacantecos thought about conflict, however, came from questioning this conceptual separation between secular and supernatural sanctioning systems. When I was given the opportunity to study witchcraft beliefs in-depth, I came to realize that it made no sense—at least for Zinacantan—to distinguish "law" from "religion" and "witchcraft."

I was given the opportunity to explore witchcraft beliefs by another member of the Harvard Chiapas Project, sociologist Francesca Cancian, who was analyzing the relationship between Zinacanteco norms and behavior (Cancian 1975). As part of her study, she developed a list of norms by employing three literate informants to complete four sentences that began, in translation, "Mariano is good (bad, respected, not respected) because . . . ." The informants were instructed to go through the Tzotzil dictionary and to construct meaningful sentence endings from as many entries as they could. The result was a vast number of Tzotzil sentences, each written on a separate notecard. After she translated and sorted these cards, Francesca Cancian lent me the ones she thought might have to do with witchcraft. There were 192. I immediately recognized some of the sentences as referring to witchcraft, such as those reporting that "Mariano is bad because he sells souls to the Earth" or "Mariano is bad because he makes others sick by coughing on them." Other sentences seemed more ambiguous, such as those reporting that "Mariano is not respected because he cries a lot." And I eventually found that a few of the sentences bore no relation to what I would have classified as witchcraft.

I used these 192 cards as the basis for interviewing three informants, two of whom had participated in Francesca Cancian's project. I asked each informant in separate interviews to sort the cards into piles reflecting similar actions, letting each man decide on the number of piles he wanted to make. After the cards were sorted, I asked each informant to explain his reasons for the groupings. The interviews were informal and exploratory, and tended to last several days. I found that all three informants were very reluctant to talk about witchcraft beliefs at first, but that they gradually warmed to the task. Their reluctance to talk about witchcraft, however, convinced me that I would have learned very little from trying to interview people without the note cards to stimulate conversations. Nor could I have obtained the same quality of information from collecting and analyzing witchcraft cases. Only long residence in several Zinacanteco households, or having the luck to find more than one "witch" willing to talk with me, could have provided information of comparable quality and quantity to that I obtained in a short time using the note cards.

Although the three informants sorted the cards into different numbers of piles and had different ideas about the prevalence of witchcraft, the overlap in their sortings suggested that Zinacantecos tended to distinguish at least three types of "witches" on the basis of the techniques they used. The first was a jealous or vengeful person who took action to make another ill. This is the kind of person Zinacantecos commonly meant when they used the term "*h'ak'chamel*" (giver of sickness). Such a witch might slice candles into little pieces to "cut" the victim's life/luck. The most frequently reported action, however, was *chonel ta balamil* (selling to the Earth), in which the witch went either alone or with a shaman to ask the powerful Lord of the Earth to take someone's soul. In a variant of this technique, a greedy person might ask for money from the Earth Lord, causing an epidemic when the Earth Lord took many souls in payment.

The second type of "witch" was someone who could cause sickness through his or her person. The line between humans who could make another sick by coughing on them or by looking at them (evil eye) and supernatural beings, such as demons or rainbows who caused illness in humans who encountered them, was not very clear. Some of the Zinacantecos I spoke with doubted that there were any humans who had the personal ability to make others sick. But even these doubting individuals seemed to believe demons could assume human form to harm living people.

The third type of person who caused illness might or might not be classified as an *h'ak'chamel*. This was a person whose "heart" was so angry from having been wronged by another that the heart "cried out" spontaneously—often without the wronged individual's knowledge or intention—to the ancestor gods of the upper world for justice. Because Zinacantecos believed that the ancestor gods acted on their own anyway to punish wrongdoers, someone who "cried for illness" could be seen as simply ensuring that proper order was restored. On the other hand, at least one of the informants I interviewed thought that vengeful people could "cry out" deliberately, in order to trick the ancestor gods into sending *ok'itabil chamel* (crying sickness) to an innocent victim. It is also true that the most common characterization of a witch in Zinacantan was as someone who "does not know how to forgive"—whose "heart" continues to seek vengeance rather than "ending" its "anger."

Because all of the major witchcraft techniques involved harming a victim's soul, I found that I could not understand Zinacanteco ideas about witchcraft without investigating their ideas about souls. Witchcraft, of course, is usually defined in terms of techniques that harm the body through injuring a person's essence. But soul beliefs turn out to be particularly important in Zinacantan because, as other members of the Harvard Chiapas project discovered, Zinacantecos attributed all illnesses and misfortunes—except disabilities resulting from obvious accidents, such as fractured bones or wounds—to "various supernatural events acting on one or both of the victim's two souls" (Fabrega and Silver 1973: 85).

Project members who interviewed shamans about illness and curing commonly portrayed Zinacantecos as distinguishing two types of souls, one an animal spirit companion (*chanul*) who lived apart from the body, and another inside the body (*ch'ulel*) made up of thirteen detachable parts. Animal spirit companions lived in a corral inside the highest sacred mountain, tended by helpers of the ancestral gods. If the animals escaped, or were let out of the corral on orders from the ancestor gods, they could be wounded or killed, causing their human counterparts to suffer similar fates. The thirteen-part *ch'ulel* was supposed to stay inside the body, but it took time for a *ch'ulel* to become accustomed to its host, making babies particularly vulnerable to soul loss. The souls of adults, while more firmly fixed, tended to wander when their bodily hosts were asleep or drunk. A person could also lose parts of the *ch'ulel* to the Earth Lord if frightened or if "sold" to the Earth Lord by a witch. Should wandering pieces of the ch'ulel not return to an awake and sober body, that body could fall ill and die (see Vogt 1965, 1969).

The Zinacantecos I interviewed about witchcraft beliefs seemed less certain about the nature of souls than the shamans interviewed by other project members. But they did all talk and act as if souls (of some kind) had a life apart from bodies. Not only did most witchcraft techniques involve harming a victim's soul, but witches were often portrayed as sending their own souls out to harm the souls of others. The animal spirit companion of a witch, for example, might attack and eat the animal spirit companion of a victim, causing the victim to sicken and die. Or witches might send their inner souls out at night to catch the wandering souls of sleeping or drunken victims, who would then wake up ill the next day.

The three informants I interviewed also seemed to believe that souls differed in their power and vulnerability. They all agreed, for example, that people with weak or poorly attached souls, such as little children, were vulnerable to illness, whereas illness was less likely to strike those with strong souls—who had fierce animals, such as jaguars, for animal spirit companions and a powerful *ch'ulel*. These beliefs were reinforced by the fact that infant and child mortality was (and is) very high in Zinacantan; nearly 40 percent of those born die before the age of 15. These beliefs were also supported by post hoc reasoning: Those who died young must have had weak souls, whereas those who live into old age must have strong ones. The three informants disagreed, however, on the vulnerability of people with strong souls. One self-confident man felt that people with strong souls were invulnerable to illness unless they committed some offense that angered the ancestor gods or unless they let down their guard by getting drunk. The most fearful of the three informants, in contrast, thought that even sober people with strong souls could be snatched by the Earth Lord, or fall ill when a clever witch tricked the ancestor gods into punishing an innocent victim.

This analysis of witchcraft and soul beliefs offered me a new way to understand the Tzotzil legal terms I had been collecting. Instead of trying to develop English translations—albeit complex ones—for Tzotzil words, I turned to thinking about the cultural logic within which such words made sense. Witchcraft beliefs enabled me to explore the "kind of world" presupposed by Zinacanteco ways of talking about conflict. I soon discovered that it was a conceptual world in which the most important and consequential relationships between people took place on the invisible level of souls.

I realized, for example, that the Tzotzil word "*mulil*"—which I had used the dictionary to translate as "guilt," "crime," "sin," and "blame"—had to be understood less in terms of a list of English words than as any action that could bring down the wrath of the ancestor gods, causing them to send sickness to an offender's home. I already knew from my analysis of cases that Zinacantecos were more interested in negotiating relationships than in punishing offenders. But I suddenly understood why they had to help disputants agree on the relationship between them. It was not out of some abstract belief in the value of reconciliation. Rather, Zinacantecos had

to agree on relationships because they lived in an imagined world where the ancestor gods were either already punishing wrongdoers or would soon do so. Children
were about to die. The parties to a dispute had to come to a mutually acceptable
solution. As along as any one person's heart remained dissatisfied, that person's angry
heart would cry out to the ancestor gods for vengeance. No wonder Zinacantecos
feared people who did "not know how to forgive."

My analysis of witchcraft beliefs also helped me to understand why it might make
sense for Zinacantecos to use the same linguistic term both for the person who starts
a fight and for the one who takes a dispute to court to try to settle it. Both people
performed similar actions in bringing a dispute into the open. Although taking a case
to court might seem to be a very different way of publicizing a dispute than attacking one's enemy or screaming insults outside an enemy's house at night, the actions
were similar in that the dispute was assumed to exist prior to the act that brought
it into the open. Zinacanteco cultural logic seemed to run something like this: Why
would someone harm someone else, or take someone to court, unless trying to
redress a previous wrong? Zinacantecos did believe that some people were bad. They
used the Spanish loan word "*manya*"[2] to refer to the inner quality that explained why
some people might be more violent, greedy or vengeful than would be expected
given the past history of their relations with those they harmed. But *manya* was a fall-
back concept. It was brought in to explain people's actions only when those actions
seemed too extreme to be explained by righteous anger or drunken stupor.

It is also true that Zinacanteco procedures for managing disputes tended to
confirm the assumption that every dispute has a long history. Because a *mulil* could
not be considered "ended" until all parties expressed satisfaction with a suggested
settlement, those who were accused of wrongdoing were given ample opportunity to
explain their actions. All defendants except the dumbest could usually come up with
some story to explain why they might have good reason to harm those who accused
them. Husbands charged with wife-beating, for example, commonly blamed their
wives for sloppy housekeeping; wives who had run home to mother accused their
husbands of beating them, people accused of theft claimed to be collecting an unpaid
loan, those caught performing witchcraft claimed to be returning sickness to the
original witch, and so forth. Such justifications seldom excused a wrongdoer. Rather,
they commonly led to long discussions about underlying problems.

When I wrote my dissertation on "Zinacanteco Law" during 1969–1970, and
later revised it for publication (1973), I organized my presentation according to the
theoretical framework proposed by Laura Nader in her introductions to the two
volumes of conference papers she edited (1965, 1969). Whereas most earlier studies
in the anthropology of law had focused on how men in authority decided cases
("dispute processing"), she proposed to focus instead on how litigants sought remedies for their problems ("the disputing process"). Nader's approach appealed to me.
It made sense that judges (and other types of third parties) could decide only on the
cases that actually reached them. The reasoning of judges—their "style" of seeking
solutions—had to be influenced by the kinds of problems that disputants brought to
them (Nader 1969).

Nader's proposed shift in focus from judges to litigants also fit with Fredrik
Barth's recommendation that anthropologists shift from studying social institutions

to studying individuals as strategic actors. Barth argued that the patterns of behavior anthropologists had been interpreting as evidence of social structure could be better understood as the outcome of innumerable decisions made by individuals operating under particular constraints and incentives (Barth 1966: 2). A focus on individual choices, he argued, offered a way to explain both stability and change. Societies appeared to have stable structures when individuals tended to make the same choices over time. But societies could change rapidly if strategic actors decided to make different choices. Barth thus offered a way of overcoming what many anthropologists had perceived to be the major weakness of British structural functionalism: its inability to account for change except as social breakdown.

Following Nader and Barth, I thus analyzed "Zinacanteco law" in terms of the constraints and incentives available to individuals involved in disputes. When I rewrote my dissertation as a book, I divided it into three sections. In the first, I analyzed the various ways of "ending a mulil" that were open to Zinacanteco disputants. Beginning with the simplest method, in which an offender went to beg pardon directly from the offended person, and moving through the various forums available to Zinacantecos for handling their disputes, I explored the relationship between types of conflicts and the methods used to handle them. In the middle section of the book I concentrated on the conceptual and practical tools available to Zinacantecos for fighting with one another. This was where I put my analysis of the words Zinacantecos used to talk about the disputing process, including my analysis of witchcraft beliefs, as well as chapters analyzing the kinds of actions that Zinacantecos interpreted as indicating aggression, such as slicing candles into little pieces, shouting insults outside someone's house at night, or going after someone with a machete or gun. In the third section, I discussed the issues that Zinacantecos could fight about, focusing on types of relationships. I followed my Zinacanteco informants in assuming that if kinsmen fought, it must be over inheritance; if spouses fought, it must be because someone had not fulfilled marital obligations; or if neighbors fought, it must be because their proximity brought them into conflict over such things as wandering animals, broken fences, or drunken insults.

I borrowed my definition of "law" from Barkun because his vision of law as a language for conducting and resolving conflicts fit with my focus on the resources available to disputants. "Law," Barkun proposed, "is that system of manipulable symbols that functions as a representation, as a model of social structure" (Barkun 1968: 92). As a symbol system, law "is a means of conceptualizing and managing the social environment" (Barkun 1968: 151). Barkun's definition also appealed to me because my study of witchcraft beliefs had led me to understand such beliefs not as a static set of assertions about the nature of reality, but rather as a flexible set of concepts that people could use to explain—and therefore to manage—almost anything that happened to anyone.

Although my analysis of witchcraft beliefs led me to the same conclusion that Carol Greenhouse reached in her senior honors thesis on "Litigant Choice" in Zinacantan, which was that "religion" and "law" are not separable domains (Greenhouse 1971: 98), I found myself uncomfortable with her assumption that religion and law formed "one internalized system of social control" (Greenhouse 1971: 3). At the time, I attributed my unease about "social control" to the fact that

I preferred to treat Zinacantecos as strategic actors rather than as potential wrong-doers. But I gradually came to realize that my unease about "social control" had deeper roots. In the rest of this chapter, I will briefly outline how my analysis of witchcraft beliefs shaped my participation in the theoretical shifts that took place in North American anthropology in the 1970s and early 1980s, and eventually led me to the conclusion reached by Marilyn Strathern (1985): that the concept of "social control" presumes a Western world view.

At the time I did my dissertation research in Zinacantan, I was not aware of the theoretical distinction between consensus and conflict models of society. Marxism had not been part of my education. I did my field work assuming that conflict was bad and its resolution good. I believed that everyone benefited from the maintenance of social order and suffered if social order was disrupted. My introduction to Marxist theory and my participation in the feminist movement, however, led me to rethink this theoretical framework. Returning to the materials I had collected in Zinacantan, I realized that the legal concepts and witchcraft beliefs I had analyzed as available to disputants in general were, in fact, more available to some than to others. Barkun's definition of "law" as a "model of social structure" was still useful, but the social structure that was being modeled was a structure of inequality.

In Zinacantan, "social structure" was biased in favor of older men who had the wealth to serve in many religious posts. Not only were such elders credited with having the strong souls needed to protect (or to harm) weaker souls, but their experience of serving the gods gave them the right to speak as representatives of supernatural authorities. By talking in ritual couplets and mouthing platitudes about how people should behave toward one another, such elders enjoyed far more power than others to determine the meaning—and therefore the consequences—of events.

More important, however, I realized that for young men and women, doing good and doing well did not coincide. To the degree that young people were "good," they remained subordinated to elders. Only by committing a wrong—such as when a young husband beat his wife, or a young wife ran away from her husband—could young couples set in motion the settlement procedures that might lead to their being allowed to set up an independent household. For older people in Zinacantan, in contrast, doing good and doing well did coincide. An elder who "generously" helped a younger person "end a mulil" could expect that young person's labor and loyalty in return.

The realization that doing good and doing well tended to coincide only for those already enjoying privileged positions helped me to understand some of my unease with the concept of social control. It meant that control was enforced mainly on those who were oppressed by the existing order. While it might be true that privileged people had to restrain themselves from committing crimes, it was also true that privileged people had little incentive to misbehave. In contrast, those on the bottom of the social order—for whom doing good meant submitting to domination—had strong incentives to disobey the norms that constrained them.

My shift from a consensus to a conflict model of society led me to drop the concept "social control" from my theoretical toolkit, but it was my interest in language that eventually led me to understand "social control" as a Western concept. Following Bohannan (1957), I had analyzed key Tzotzil terms in order to explore how Zinacanteco disputants and third parties understood what they were doing. I thus found it easy to fit my analysis of Zinacanteco disputing processes into the theoretical framework being developed by Clifford Geertz (1973), as I learned about his work from my colleague at Stanford, Michelle Z. Rosaldo. I liked Geertz's vision that "culture" is embodied in publicly shared symbols rather than as rules inside the heads of individuals. I had never been comfortable with the ethnoscientific approach that other ethnographers had used to study law in Chiapas (Black and Metzger 1965). Although I believe that formal elic-itation techniques can be helpful for mapping restricted cognitive domains,[3] I doubted their usefulness for grasping the complexities of large and ill-defined ones, such as how a people might think about conflict and its management. It is also true that I was unable to use such techniques when doing research in the 1960s. As a young woman, I lacked the social power to compel the important Zinacanteco men I interviewed to take me seriously when I asked them to answer questions that even I found simplistic, boring, and repetitive. One wit, for example, answered my repeated requests to "tell me the name of another kind of mulil" by listing, as a separate kind of theft, every moveable object in a Zinacanteco household.

I also liked Geertz's emphasis on understanding the actor's point of view. It fit with Barth's directive to study the choices made by individuals, but Geertz's hermeneutic approach provided a more satisfying way of understanding individual choices than the market model of maximizing individuals assumed by Barth (1966)—and devastatingly criticized from a Marxist perspective by Talal Asad (1972). Geertz's interpretive methodology required paying careful attention to what actors said they were doing. Rather than simply assuming that actors were trying to maximize utilities or that their stated preferences could be taken at face value, Geertz advocated exploring actors' unstated assumptions about the world they lived in. His framework thus encouraged me to continue my practice of trying to understand Tzotzil terms by exploring the imagined world in which they might make sense.

Because I preferred a conflict model of society to a consensus one, I used Geertz's interpretive approach to explore how people experienced and enacted social inequal-ity. I focused on trying to understand the unstated assumptions behind people's prescriptions for how to become a respected person in one's social group—or at least for how to avoid becoming someone whom others could safely ignore or treat with disdain. And because I assumed that people derived their assumptions about the social world from their experiences of living in it, I explored the social processes responsible for allocating wealth, prestige, and power.

By analyzing the organization of social inequality in Zinacantan as it operated in the 1960s, I came to understand people's concern with "ending the anger in people's hearts" as integral to a system of stratification in which control over labor was more important than control over capital, and in which bachelors' need for bridewealth encouraged them to work for elders who could help them marry. Zinacanteco elders may have acquired respect and influence by serving in religious posts, but in order to obtain the wealth needed to take a religious post, a man had to accumulate many working juniors in his household and many debtors in his community. No wonder influential men in Zinacantan preferred "reconciliation" to punishing wrongdoers. They needed people to willingly fulfill their obligations.

My interest in exploring the relationship between ways of organizing social inequality and methods of managing disputes led me to study another Chiapas community in which I suspected that access to capital was more important than control over family labor for achieving wealth, power, and prestige (Collier 1979). San Felipe, which is located between Zinacantan and the city of San Cristobal, did turn out to be a community where the wealthiest men owned trucks or small businesses, or held salaried positions. It was also a community where young people married for "love," and where men who had capital avoided serving in religious posts, preferring instead to host private celebrations for guests of their own social class. Not surprisingly, I found that the local judge interpreted his job as one of upholding laws by punishing wrongdoers rather than as one of trying to calm angry hearts. In a community where the wealthiest and most influential men had little interest in forging or maintaining relationships with neighbors and distant kin, but instead wanted the freedom to deploy their capital for personal advantage, the local judge faced the task of ensuring that individuals did not harm one another.

My comparison of how third parties in Zinacantan and San Felipe handled trouble cases led me to rethink my understanding of the dispute between Gluckman and Bohannan over how to analyze judicial processes (1969). Whereas I had earlier adopted Bohannan's view that judicial processes are culturally specific, I came instead to sympathize with Gluckman's position that people everywhere, when faced with a similar task, tend to come up with similar concepts for managing it. But instead of assuming with Gluckman that "judges" everywhere face the same task (that of assessing events in terms of social norms, albeit varying their solutions according to the degree of societal complexity or the type of productive technology), I assumed that the task imposed on third parties varied according to the social organization of inequality.

As I worked on developing ideal-typic models for analyzing the relationship between ways of organizing social inequality and disputing processes (1988), I found that the English words available to me for writing about inequality were inadequate to express my growing understanding of different social worlds. I thus started asking of English words the same question I had asked earlier about Tzotzil ones: In what kind of an imagined world does this concept make sense? Marilyn Strathern put into words what I was coming to perceive, when she wrote that concepts such as "social control" draw "on a model of social life and human behavior which belongs very much to the industrial west, as well as state systems of government"—it is a model based on the idea that "order is the proper state of society, and (that) society itself is imposed upon individuals who are by natural propensity asocial beings" (Strathern 1985: 113).

Moreover, Strathern's criticism of the western idea that "behavior either acts out relationships or disrupts them" (Strathern 1985: 122) helped me to rethink my understanding of what Zinacantecos were doing in the 1960s. Like the anthropologists Strathern criticizes, I had thought about Zinacantecos as concerned with "repairing" relationships. But I came to accept Strathern's argument that "reparation" is the wrong word because it implies the notion of something broken that can be repaired. In Mt. Hagen, New Guinea where she worked, disputants were not "repairing" previously existing ties. Rather, they were publicly negotiating the kind of relationship that existed between them. What "looks like reparation," Strathern argues, is "better understood as a public evaluation of relationships" (Strathern 1985: 124). This seemed a much better way to understand what I had observed Zinacantecos doing in the 1960s. Like the people of Mt. Hagen, Zinacantecos were continually negotiating their relationships because they lived in a social world where people got ahead by shedding obligations to others while trying to obligate others to them.

This insight led me to develop a new understanding of Carol Greenhouse's observation that Zinacantecos procedures for handling disputes and for curing illness were part of the same system. It was not a system of social control as she had suggested; rather, it was a system for negotiating relationships. In the 1960s, Zinacantecos, like the people of Mt. Hagen studied by Strathern, lacked a context for inventing a "description of society as norm-bound in the interests of peaceful coexistence" (Strathern 1985: 128). When Zinacanteco elders and shamans mouthed platitudes about how people in particular relationships should behave toward one another, they were not commenting on behavior from outside the dispute, but rather participating in the disputing process itself, helping disputants to (re)negotiate their relationship. As Strathern observes, it is a western notion that "adjudication" does "not itself participate in the actions under scrutiny" (Strathern 1985: 128).

In summary, I discarded the notion of "social control" because I found that trying to analyze Zinacanteco disputing processes as a form of "social control" had the effect of distorting what I had come to understand the Zinacantecos as doing. Like the people of Mt. Hagen studied by Marilyn Strathern, Zinacantecos in the 1960s seemed to have no conception of a living in a "society," much less one so fragile that it could continue to exist only so long as individuals were restrained from acting on their natural selfish desires. Instead of experiencing their world as one in which individuals needed to be controlled, Zinacantecos seemed to experience their world as one in which individuals were continually negotiating their relationships with others. Theirs was a world always in process. Most negotiations between people occurred peacefully, but sometimes people fought with one another, causing anger in the hearts of some individuals that cried out to the gods for vengeance. Should this happen, people had to find a way to calm angry hearts before children started dying. Disputants could negotiate directly with one another, or ask a respected elder to help them find a solution to their problem. But should an elder participate, that elder was in no sense a judge who stood outside the conflict. Rather, respected elders were participants along with the disputants in seeking solutions to calm angry hearts. No one involved in the disputing process was trying to preserve social order by punishing those who violated accepted norms. Their ideal world was not one in which a fear of punishment kept individuals under "control." It was instead one of "happy hearts."

Finally, as I worked with June Starr in thinking about how world historical processes have affected the organization of social inequality and disputing processes in local places (Starr and Collier 1987, 1989), I began to wonder what had happened in Zinacantan since the Mexican oil boom of the 1970s and the subsequent debt crisis of the 1980s. George Collier's work suggested that Zinacantan had emerged from these economic upheavals as a community where control over capital is more important than control over labor (Collier 1994). I thus decided to return to Zinacantan to restudy disputing processes. Although I have not yet had a chance to analyze the data I collected between 1997 and 1999, I think that the economic changes have had less effect on Zinacanteco legal concepts and behavior than I had previously thought. The Zinacanteco judges I observed during the autumns of 1997 and 1998 were still less interested in enforcing rules than in helping disputants to negotiate a relationship that would allow them to live together in peace. The economic shift from labor- to capital-intensive agriculture, however, does appear to have affected the willingness of important men to help their kin and neighbors settle quarrels. No longer are respected elders willing to serve as mediators. Instead, only elected and appointed municipal officials are available to perform this task. The Zinacantecos have also added the word "*kostumbre*," borrowed from the Spanish word for "custom," to their vocabulary to talk about their practices that differ from those of most Mexicans. Through their participation in national and international movements for indigenous rights, the Zinacantecos are developing new ways to assert the political and legal autonomy that was previously theirs because of neglect by the Mexican state. Judges, in particular, have become vocal advocates of their right to administer justice according to Zinacanteco "customs," in response to increasing pressures from the national and state governments for judges to follow codified laws.[4]

Notes

I would like to thank George Collier, June Starr, and Mark Goodale for their comments on this chapter. The recent research in Zinacantan and the writing of this chapter were supported by the National Science Foundation, grant SBR-97-10396, "Mapping Interlegality in Chiapas, Mexico." The earlier research in the 1960s was supported by National Science Foundation predoctoral and postdoctoral fellowships.

1. Although I found it difficult to hang out at the town hall, I did have access to the field notes of other members of the Harvard Chiapas Project, such as three young men who were able to hang out and joke with town hall officials in Zinacantan (Young 1965) and the neighboring Tzotzil community of Chamula (Prokosch 1964; Rus 1973).
2. Although the Tzotzil word "manya" might appear identical to the Spanish word "maña," from which it is borrowed, the Tzotzil word is accented on the last syllable, as are all Tzotzil words, rather than on the first syllable. Moreover, the Tzotzil word can assume the possessive prefixes and suffixes common to other Tzotzil nouns.
3. I have long admired James Spradley's use of formal elicitation techniques, as described in his book *You Owe Yourself a Drunk* (1970). I have often recommended this book to others.
4. As of July 1999, the state government of Chiapas passed a new law granting indigenous communities the right to administer justice according to their own "usos y costumbres" (roughly, practices and customs). But because the law allows indigenous courts to practice only those "customs" that do not violate internationally recognized human rights or the rights guaranteed to individuals by the Mexican Constitution, the state is able to use a discourse of human rights to require indigenous judges to follow codified laws. State officials, for example, seem to be particularly fond of Article 7 of the Universal Declaration of Human Rights, which states that "All are equal before the law and are entitled without any discrimination to equal protection of the law."

References

Asad, Talal
    1972   Market Model, Class Structure and Consent: A Reconsideration of Swat Political Organization. Man 7(1): 74–94.
Barkun, Michael
    1968   Law Without Sanctions: Order in Primitive Societies and the World Community. New Haven, CT: Yale University Press.

Barth, Fredrik
    1966   Models of Social Organization. Royal Anthropological Institute Occasional Papers 23.
Black, Mary and Duane Metzger
    1965   Ethnographic Description and the Study of Law. American Anthropologist 67(6): pt. 2: 141–165.
Bohannan, Paul J.
    1957   Justice and Judgment among the Tiv. London: Oxford University Press.
Cancian, Francesca
    1975   What are Norms? New York: Cambridge University Press.
Collier, George A.
    1994   Basta! Land and the Zapatista Rebellion in Chiapas. Oakland, CA: Food First Book.
Collier, Jane F.
    1973   Law and Social Change in Zinacantan. Stanford, CA: Stanford University Press.
    1979   Stratification and Dispute Handling in Two Highland Chiapas Communities. American Ethnologist 6(2): 305–327.
    1988   Marriage and Inequality in Classless Societies. Stanford, CA: Stanford University Press.
    1999   Models of Indigenous Justice in Chiapas, Mexico. A Comparison of State and Zinacanteco Versions. PoLAR 22(1): 94–100.
Evans-Pritchard, Edward E.
    1976   Witchcraft, Oracles and Magic Among the Azande. Oxford: Oxford University Press.
Fabrega, Horacio, Jr., and Daniel Silver
    1973   Illness and Shamanistic Curing in Zinacantan. Stanford, CA: Stanford University Press.
Geertz, Clifford
    1973   The Interpretation of Cultures. New York: Basic Books.
Gluckman, Max
    1955   The Judicial Process among the Bartose of Northern Rhodesia. Manchester: Manchester University Press.
Greenhouse, Carol
    1971   Litigant Choice: Non-secular and Secular Sanctions in Zinacanteco Conflict Resolution. Unpublished Senior Honors Thesis. Harvard University.
Guiteras Holmes, Calixta
    1961   Perils of the Soul: The World View of a Tzotzil Indian. New York: Free Press of Glencoe.
Hermitte, Esther M.
    1964   Supernatural Power and Social Control in a Modern Maya Village. Unpublished Doctoral dissertation in Anthropology. University of Chicago.
Hoebel, E. Adamson
    1954   The Law of Primitive Man. Cambridge, MA: Harvard University Press.
Laughlin, Robert
    1975   The Great Tzotzil Dictionary of San Lorenzo Zinacantan. Washington, DC: Smithsonian Institution Press.
Malinowski, Bronislaw
    1948   Magic, Science and Religion and Other Essays. Glencoe, IL: Free Press.
Nader, Laura
    1969   "Styles of Court Procedure: To Make the Balance." In Laura Nader (ed.) Law in Culture and Society. Chicago, IL: Aldine.
Nader, Laura (ed.)
    1965   *The Ethnography of Law*. American Anthropologist (special publication) 67(6).
    1969   Law in Culture and Society. Chicago, IL: Aldine

Pospisil, Leopold
   1958   Kapauku Papuans and their Law. Yale University Publications in Anthropology 54.
Prokosch, Eric
   1964   Court Procedure in the Settlement of Disputes in Chamula. Unpublished manuscript. London School of Economics.
Rus, Jan
   1973   "One Court, Two Cultures: Rhetorical Strategy and Cultural Interference in a Changing Maya Community." Paper delivered at the 1973 meetings of the American Anthropological Association.
Spradley, James
   1970   You Owe Yourself a Drunk: An Ethnography of Urban Nomads. Boston: Little, Brown.
Starr, June and Jane F. Collier
   1987   Historical Studies of Legal Change. Current Anthropology 28(3): 367–373.
Starr, June and Jane F. Collier (eds.)
   1989   History and Power in the Study of Law. Ithaca, NY: Cornell University Press.
Strathern, Marilyn
   1985   Discovering "Social Control." Journal of Law and Society 12(2): 111–134.
Vogt, Evon Z.
   1965   Zinacanteco "souls." Man 29: 33–35.
   1969   Zinacantan: A Maya Community in the Highlands of Chiapas. Cambridge, MA: Harvard University Press.
Young, Stephen B.
   1965   "Their People's Servants:" Political Officials in a Highland Maya Community. Unpublished ms. of the Harvard Chiapas Project.

CHAPTER 5
EXPLORING LEGAL CULTURE IN LAW-AVOIDANCE SOCIETIES*

Robert L. Kidder

Sociologists tend to speak of ethnographic methods as standing in contrast to the work of our "number crunchers" who seek to test hypotheses by using samples called for by statistical models. Our embrace of ethnography (or our more general term, "field methods") has roots in a sociological tradition, "symbolic interaction," which emphasizes the relationship between perceptions of reality and actions based on those perceptions. Sociological methods such as participant observation, therefore, are justified on the grounds that one must "see the world as one's subjects see it" in order to understand the choices they make, as inferred from their actions.

Having conducted observational, or "ethnographic,"[1] research on questions of law and society in both Amish[2] and Japanese settings (see Kidder and Hostetler 1990; Kidder and Miyazawa 1993), I have encountered research issues that may stem from the fact that for both of these groups, scholars had spoken of a culture that values law-avoidance. In this chapter, I will address aspects of this kind of research that may or may not be affected by this public posture.

The problem I address in this chapter concerns what people in these two societies say about the absence of conflict in their lives and about the viability of law-avoidance as a basis for social organization. It is a problem in deciding how to define, study and report on what some have called "legal consciousness"[3] or "legal culture."[4]

I use the term "law-avoidance societies" for those societies that display three characteristics: (1) a dominant ideology that law systems, law-centered thought (such as rights consciousness) and legal confrontation are bad for society; (2) a formal legal system that is underdeveloped when compared with other late-twentieth century Western social systems; and (3) a demonstrated incidence of infrequent use of formal legal institutions by most segments of the population. A law-avoidance ideology treats rights consciousness as a social pathology.

"Law-avoidance" does not mean that people in these societies seek to violate laws or seek to overthrow governments. By most measures, the Japanese and the Amish seem to outsiders to deviate in the direction of over-conformity with most laws of the states in which they live. Rather, law-avoidance refers to ideological statements about how problems arise between people and how they should behave when such problems do arise, and to the apparent connection between this ideology and the relative absence of formal legal institutions. Nader calls this phenomenon "harmony ideology" (1990). Similar concepts appear in a number of other studies (e.g., Engel 1984; Greenhouse 1986). I prefer to identify it as law-avoidance because that term emphasizes the extent to which the ideology represents a response to the development of Western legal institutions.

Japanese and Amish societies hold similar ideas concerning the meaning, values, and norms relating to social conflict. Both ideologies define conflict as a disruption of a desirable state of relationships that all parties have an obligation to repair or re-instate. They define confrontation, adversarial behavior, and public accusation as wrong, in both a spiritual sense and a practical sense. Both societies invoke their own idea of community as the ideal object of concern to all parties. Conflict disrupts community by selfish insistence on individual interests. Good community members who find themselves involved in conflict are directed to keep the problem quiet, examine their own position for weaknesses and evidence of their own errors, and seek an accommodation that will preserve relationships and the well-being of the broader community. Self-interest must be weighed against the larger good of the group. Compromise is good. Excommunication or social shunning is an approved way to assure that individuals will comply. "Lumping it,"[5] though not spoken of in that language, is also an approved solution to conflicts.

The actual details of the Japanese and Amish ideologies differ, and stem from very different social histories. In the next two sections, I will describe the ideas of the Amish and the Japanese as I encountered them prior to my field research experiences in both societies.

Amish Legal Consciousness

The aspect of Amish ideology that we can call its legal consciousness is centered around the construction and preservation of a pure or holy church community, separate from "the world." This community is a social entity composed of imperfect, weak humans who must try their utmost to prepare the community as a pure offering to God.[6] The community, which is coterminous with the church, is a unitary entity and must take precedence over the individual. The church community, like the fertile farmland the Amish prize, is a trust which they must tend and pass on to succeeding generations. Conflict is an impurity based on human weakness that threatens the community. It must therefore be erased, washed away, or expelled.

Some of the most conspicuous Amish practices, including their rejection of modern technology and social forms, derive directly from their concerns about conflict. Their refusal to use electricity, cars, telephones or other things outsiders would call labor-saving devices, stems from a constantly challenged position that members must act to protect the community from the destabilizing side effects on human relationships that these inventions bring. For the same reasons, they reject commercial insurance (the community needs the solidarity that comes when people rely on each other rather than institutions outside the community to rebuild after catastrophe). Likewise they avoid large accumulations of funds in their churches (solidarity also arises from meeting church needs as they arise instead of planning for the future). The rejection of law fits into this pattern. Reliance on law would sap the strength that the community receives from handling problems internally.

Japanese Culture of Harmony

Kawashima's discussion of the Japanese legal system presents a social world in which an ancient Confucian concept, "Wa" (loosely translated as harmony), dominates thinking about what one should do when conflict arises (1963). Kawashima introduces this concept as a way of explaining what he accepts as fact: that Japanese people in general avoid public confrontation and much prefer nonlegal harmonizing procedures such as mediation or even "lumping it" over legalistic actions based on one's rights. Kawashima goes on to argue that because these harmonizing preferences dominate Japanese thinking, all Japanese institutions operate to strengthen steps that help people avoid confrontation. Japan's legal institutions are therefore relatively insignificant when compared to those in Western capitalist cultures. More recent versions of this thesis deal with apology (Wagatsuma and Rosett 1986) and with the social psychology of justice (Hamilton and Sanders 1992). All studies with this point of view share the purposes of documenting and accounting for the extent to which Japan has achieved advanced capitalist development without the full development of legal institutions that have seemed necessary in other similarly developed economies, especially the United States.

Problems of Reasoning and Evidence in Legal Ethnographies

The type of research that I have conducted on dispute processing in Amish and Japanese settings raises three issues related to the effort to "let people speak for themselves" in forming a sociological analysis. First, how should I interpret statements made to me by my subjects that reject the comparative analytic approach on which my research is based? Second, how can statements and actions of either the Amish or the Japanese be given reliable and valid interpretations? And third, how can my research be a true vehicle for the voice of my subjects and yet contribute to sociological knowledge?

A. The Logic of Comparison

Comparison is at the core of nearly all research and analysis done in sociology, whether its practitioners are concerned about hearing the peoples' voices or not. Macrosociological theories, such as structural-functionalism and conflict theory are couched in terms of comparison. For structural-functionalists,[7] for example, comparison reveals alternative ways of dealing with fundamental problems of social organization. The logical conclusion from such a view is that social systems must either solve those problems with some kind of social construction, or they will fail. Conflict theories in sociology[8] also suggest the value of comparative research as a tool for mapping the different ways that fundamental social contradictions turn into overt political and legal expression. Conflict theorists often treat comparative study as a way of clearing away the thicket of apparent difference between societies to expose the common processes of conflict that support the theories.

Unlike these other more macrosociological theories, Symbolic Interactionist theory redirects our attention to the interplay between cognitive processes of interpretation and social constructionist processes of consolidating those interpretations into shared notions of reality. Symbolic Interactionists, therefore, approach comparative research as an opportunity to confirm their theoretical stance. It can show how differently the situations can look to peoples with a different world view and how their view relates to the social worlds they have constructed. Symbolic Interactionists say: "What you think is X may not look like X at all to Y people, and you need to study Y people along with As, Bs, Cs, etc. in order to get a better understanding of what you think is X." In the language of more recent discussions of "voice" (e.g., Taylor et al. 1995), Symbolic Interactionists, for many decades, have taken the position that all interpretation must start with the voice of the persons being studied.[9] The presumption is that this voice reflects an inner cognitive process that relates directly to observable action.

In the research I have conducted among the Amish and in Japan, I was interested in finding out how people who shun legal institutions arrange their affairs so that they can appear recognizably Amish or Japanese in the face of conflict. My epistemological approach is fundamentally comparative, and my methodology of observation derives from the tradition of Symbolic Interaction. As comparison, I treat the Amish and the Japanese as "cases"[10] of some abstractable process. For example, I assume that terms used regularly in law and society studies, such as "conflict," "formal legal institutions," and "avoidance," have meanings that can be applied across cultures. Comparative research values both what is unique about a "case" (e.g., a "case" of law-avoidance social organization), and what it has in common with other such "cases."

The problem I raise here is that the "voices" I heard in both Japan and with the Amish not only supported a conclusion about similarity of legal culture, but denied, in their own "voices," that the similarity I saw between them existed. The Amish, for example, saw their actions as stemming from their Christian commitments, and usually emphasized the differences between their actions and those of the "English" (the Amish term for outsiders). This view of outsiders included the secular, high-consumption, fast-paced Japanese, even if the Japanese do have a dim view of legal institutions. Japanese people who have heard me compare their legal consciousness and law-related actions to the Amish have objected on the grounds that the Amish are "so different" from themselves.

This may seem like a mere quibble. Sociologists have been moving their field observations into more general theoretical abstractions as an essential part of what doing sociology means. Of course people who are not themselves doing sociology do not routinely engage in the kind of thinking that sociologists do. But what should we do if the people about whom we are speaking reject the abstractions we develop? How much authority should their voices have? What place is there for sociological abstraction?

B. The Question of "Voice": Whose is it and What Does it Mean?
Since both the Japanese and the Amish treat legal confrontation and adversarial behavior as deviant, one problem that confronted me in studying the relationship between these groups and the legal institutions having jurisdiction over them concerned weighing their *voices* against independent evidence about their *actions*. Wolcott nicely captures the flavor of this problem: "Interviewing is not all that difficult, but interviewing in which people tell you how they really think about things you are interested in learning,

or how they think about the things that are important to them, is a delicate art. My working resolution to the dilemma of assessing what informants say is to recognize that they are always telling me *something*. My task is to figure out what that *something* might be" (1995: 105).

My research on the Amish exposed to me a related but somewhat different problem. My purpose was to investigate how Amish communities dealt with conflict, both with outsiders and within the community. However, for Amish persons to speak of being involved in conflict or to speak of conflict among other Amish persons, they must enter into a dialogue that reveals their own, or their community's failures. They would have to describe the details of their own deviance. If the ideology I described earlier does dominate Amish thinking, such talk itself would be deviant, or at least a sign of weakness in either the speaker or the speaker's church community. So if I hear people describing actual conflict, as I did, should I think of my informants as marginal people who are not really following the prescriptions of the ideology, or do I interpret their words as an indicator of both a dominant ideology and patterned modes of deviation from it? Even more confounding, can I interpret talk of conflict as natural among the Amish, or is it something that I forced with my questioning?

I approach these kinds of questions from a "presentation of self" (Goffman 1959) perspective on interpreting conversations, actions, and what is said in interviews. I make no assumptions about the true state of peoples' meaning or their true "voice." Rather, I assume that their words and actions contain information about what they want the observer to think about them. Words and actions may also contain additional information that can feed an observer's interpretations. However, these added interpretations are likely to go beyond the speaker's "voice." I turn now, therefore, to the question of "voice" as it may affect studies of law-avoidance societies.

C. Sociological vs. Subject "Voices": Am I Imposing a Conceptual Framework?

Given the ideals expressed by Amish and Japanese in general, my approach to studying dispute management runs the risk of imposing a conceptual framework that would produce distorted conclusions from their point of view. If their ideal is to avoid conflict with others, if they see what I would call conflict (having external characteristics and involving forces operating at a social level) as a personal weakness, then I face a problem of interpretation. If someone tells me that "we have no conflicts," I first must ask: Am I hearing an honest expression (i.e., one that the speaker does not consider to be an outright lie or deception) of the speaker's view of normal, everyday life, or am I getting a "tour bus" version of what being Amish means? Even if I think it is an honest statement, my sociological analysis may give both the speaker's words and related actions an interpretation she/he would not make.

Examples: Facing Issues in the Field

For example, what have I learned when I hear an Amish man saying that he and his family would probably leave their homes and migrate elsewhere if the State of Pennsylvania Highway Department went ahead with plans for a four-lane highway through their community? Does this man see migration as his only real choice? Or is this description crafted with other purposes in mind.

One interpretation of the statement about moving because of the highway could be that this is a fine example of the Amish ideal of "turning the other cheek." Here is a man who is prepared to uproot his family and seek more congenial conditions elsewhere rather than stand and fight for his rights. He may harbor private thoughts of resentment about this, or even an urge to fight the highway, but he is conforming his statements and actions to the tradition he has chosen to follow. This man is saying something he would not hesitate to say in front of his whole church congregation. Sure enough, the Amish avoid lawyers and courts. I can treat this as a case of avoidance. Or can I?

At least two other interpretations are plausible. One is that real estate prices have become so inflated in Lancaster County, because of the westward expansion of the Philadelphia metropolitan area, that this farmer can sell his farm for a huge gain and establish a larger, more profitable farmstead further from the urban sprawl. Do his words convey an attitude toward conflict or a rationalization, coated with Amish ideology, for making a calculated business decision? By stating his thoughts the way he has, he could be disguising a different, less righteous line of thought and action.

The second possibility is that this Amish man sees me as a possible conduit to those who make decisions about matters such as highway building. Later in my research, I learned that State of Pennsylvania officials view the Amish as a major source of tourist income. During the late 1980s, for example, the state calculated that overall tourism generated 400 million dollars of income, and of that 200 million was directly attributable to the fact that people want to come and see the Amish. Just by being Amish, this group brings in more state revenue than the Liberty Bell and Independence Hall combined. I also heard Amish leaders expressing detailed awareness of their importance to the state. They send delegations to Harrisburg regularly to confer with state officials whose responsibilities impinge in some way on Amish life. They know that these officials face strong political pressure to keep the Amish in Pennsylvania, and this means keeping them happy enough to stay put. Though they almost never vote in elections, and never run for public office, the Amish exert substantial political pressure, at least within the state of Pennsylvania.

Therefore, when the Amish man tells me that he may have to just "move the family," this could be a sincere expression[11] of his way of thinking, or it could be intended as a signal, through me and what I might report from my research, to state authorities: "Scrap the highway or you will lose a golden goose."

Amish communities often take principled religious stands that put them in direct violation of state or national laws. Some of these conflicts are the result of Amish moving into jurisdictions where existing laws contradict their principles. For example, after years of refusing to send their children to school beyond the eighth grade in Wisconsin, the Amish were granted the right to their own standards for education[12] by the United States Supreme Court. Later, as Amish communities expanded into other states, local officials in many of these states went through the same paces of demanding full compliance with state school-attendance laws and finally working out a compromise that allowed the Amish to control their own schools. Other conflicts arise when the state initiates a new law or regulation that impinges on

Amish life. Amish in Minnesota refused to mount triangular reflectors on the backs of their horse-drawn buggies, though the state law mandating them was designed for their own safety. They concluded that such reflectors were decorative and therefore violated church doctrine. Finally, some conflicts occur because of the introduction of new technologies in the general population or as the result of political initiatives that demand the compliance of "all good Americans" but violate Amish beliefs (e.g., the introduction of animal waste runoff controls to Amish farms so that rainwater going through their manure piles would not pollute the Chesapeake Bay).[13]

However, when Amish leaders involved in handling these situations talked about them, their identity as conflicts dissolved into a different Amish reality. Amish leaders told us[14] how they would go to top state or national decision-makers (e.g., the head of Pennsylvania's Department of Education, or General Hershey, director of the Selective Service System during the Vietnam War) and "work things out" with them. Instead of confrontation, they said, they "reasoned together" (a good Biblical expression) with the "folks" in those offices. They would bring the decision-makers some fresh homemade Amish bread or fresh vegetables from the garden, talk about "this and that" and reach an agreeable solution to an unacceptable regulation that government sought to impose on them. In these descriptions the narrator seems to be saying "There is no conflict here, just misunderstanding, and if we are reasonable and give people the chance to think about things, they will do the right thing (i.e., make decisions in our favor)."

If a local draft board, for example, was demanding that an Amish boy join the army, an Amish leader would just "call up my good friend, General Hershey" and describe the problem. A half-hour later, he says, he would get a call back from General Hershey saying the problem had been solved—the local board would grant conscientious-objector status to the boy. He knew he could do this because he had been to General Hershey's office many times in Washington, and General Hershey had come and sat with him on the porch of his farm house while they talked about "the situation."

I have at least two ways of interpreting what I hear in these interviews. I can look at these actions as examples of good old-fashioned ward politics, "schmoozing," the distribution of political favors to a preferred constituency. If I do this, my study fits nicely into a law and society literature about the "law in action" in situations involving conflicts that are resolved "in the shadow of the law" but not involving formal legal actions. On the other hand, if I make this interpretation, I am negating what the Amish say about themselves, and perhaps I am imposing a conceptual framework on them that distorts the world from their point of view. As a sociologist, is it my obligation to present the "voice" of these people (in which case, I serve primarily as a hopefully sensitive reporter of what is said), or do I fit their words and actions into a sociological framework that they would find alien (and perhaps even insulting)?

If I do impose a conceptual framework, then what has become of my claim that I am doing sociology ethnographically, which, as I have already explained, is based on the notion that people act on the basis of the way they define reality? To state this dilemma in terms of research strategies, do I interpret the statements I hear about the nonexistence of conflict as an expression of an Amish world view that shapes the actions of the speaker, or do I see those statements as akin to the crafty language of a Chicago ward, boss who maintains his power by disguising every aspect of its under-lying sources?

For another Amish example, I turn to the ways Amish people talked with me about what I would call conflict within the community. I heard people describe intracom-munity situations that sounded like conflict to me. But they were telling them to me as examples in which a person's human weaknesses were being put to the test.

One Amish man who had developed a business selling textbooks for Amish schools was widely suspected of having manipulated the minimalist, egalitarian, and democratic structures of various Amish communities to insure that he would have a virtual monopoly on the textbook business in their one-room schools. Others I interviewed brought these activities up and clearly disapproved of his behavior. Yet even an Amish school board member who was trying to end this monopoly and introduce what he considered to be better books talked of the problem as one involv-ing the "questionable" Amishness of the other, a view that I heard from several other interviewees. While they viewed his actions as harmful to the community, they described it as the result of his personal straying from Amish ways of thinking and acting. They also predicted that he would "go too far one day" and get himself excommunicated.

From my point of view these conflict cases within the community were elaborate instances of long-simmering tensions involving complex networks of relationships. I had found, and reported on, networks and tensions much like these in research I conducted in India (Kidder 1973). But in the Indian case, I found that multiplex relationships of this type regularly involved law courts as arenas for the interplay of conflict and cooperation over extended periods of time. Hence, I felt a sense of recognition when I discovered long-term complex conflicts in Amish communities.

For example, the monopolistic bookseller described those who wanted to change the school books as being inappropriately pushy and trying to change things too fast. But he was also weakened in his position by the fact that he had lost a book manu-script submitted to him (he was also a publisher of Amish books) by another Amish family on behalf of the deceased author (their father). The manuscript went into detail about the role of several Amish elders in handling various situations such as those described above. The author's children suspected that the bookseller had purposely "lost" the manuscript so that he could delay its publication until he wrote his own version of those events. His status was further weakened by the fact that his role in dealing with those situations had made him spend much time away from farming, traveling to distant places, and spending a lot of time with people of "the world." Because of these activities, others in his community described him as flirting with excommunication.

I listened to numerous descriptions of long-running, highly complex "conflicts" (my word) like the above that pitted people against each other over many years, some-times decades. I heard of cases of revenge, Amish style. I heard of conflicts so bad that churches were actually split. And a recurrent theme in many descriptions was how conflicts led to whole families leaving the church entirely. Yet in telling of these situations, Amish informants would speak of them as lapses into the all-to-human weaknesses of *hochmut* (pride) or jealousy, requiring personal growth on the part of what Indian litigants would have called "my opponent."

I went into this research seeking comparative information about ways of dealing with conflict. My ideas about conflict came from my own previous research and the research context of law and society literature, especially that which is informed by sociological theory. I thought the Amish would be an interesting case study of organized nonlegal conflict management. If I continue to call what I found among the Amish "conflict," how do I incorporate their "voice" into my ethnography?

Research on Japanese Litigation

A. Who Speaks for Japanese Culture?

In the case of my research in Japan,[15] my focus was on a group that had actively pursued a strategy of litigation in direct violation of the ideology of harmony. I was seeking information about my respondents' involvements with the courts and with lawyers. I was also investigating the views and actions of the lawyers who were involved on the plaintiffs' side. I was focusing, therefore, on a large group of people whom I considered to be actors in a case of conflict in Japan. I expected the conclusions to contribute to our general understanding of the relationship between legal institutions and dispute processing. I thought the Japanese cases would offer unique insights because of Japan's status as an economic giant, its well-documented restrictions on legal-system development, and debates about the significance of its unique culture as a law-avoidance force. As with the Amish, I approached the research with the presumption that these questions would be best explored by hearing the "voice" of the Japanese, and in particular of the Japanese litigants and lawyers I came to study.

One of my purposes was to investigate cases of public disputing in order to compare them with literature that advances culture-based theories of Japanese non-litigiousness and underdeveloped legal institutions. Having found culture-based theory inadequate to explain what I was observing in these cases,[16] I wondered: Were my negative conclusions about the cultural theory of Japanese harmoniousness a foregone conclusion produced by my implementation of a biased research agenda? If I focus on people who have gone against the norm, should I not expect them to hold opinions and exhibit behaviors that go against those norms?

Here are examples of the problem. Informants told stories about difficulty finding lawyers to take their cases. These stories contributed to my interpretation that nonlitigiousness stems in part from the very high costs that government policy has imposed on those who want to sue. The limited access to lawyers stems from government limits (supported by both local and national bar associations) on the size of the legal profession. A potential litigant's obstacles are increased by stiff fees imposed by the courts on those who file cases. Costs are raised even further by the glacial pace of Japanese litigation,[17] a condition created by strict Ministry of Justice limits on the numbers of lawyers and judges. I reasoned that if this group of people stayed out of litigation for nearly fifteen years primarily because it took them that long to find lawyers who would handle their cases for a price they could afford, statistical evidence of Japanese nonlitigiousness might well reflect similar levels of frustration in the Japanese population in general.[18] As my research continued I began to encounter a wider network of groups that had filed similar lawsuits all over Japan, and found that this network was connected, through activist lawyers, to an even wider network of "troublemaking" on a range of issues that went well beyond the problems of air pollution (e.g., prisoners' rights, "Bullet-train" noise pollution, returning land to Okinawans by closing American military installations).

What "voice" have I heard in listening to these people? In part, this is a sampling question: Perhaps the reason I did not hear people talking about the need for harmony and the negative personal characteristics of people who sue instead of keeping harmony, is simply that I was talking to the wrong people. Maybe these are the disaffected few, maybe they are not "typical Japanese," maybe something went wrong with their acculturation. How, for example, would they have answered the questions put by Hamilton and Sanders (1992) in their surveys?[19]

Perhaps I heard only the "voice" of the people I interviewed and observed. Perhaps such a study cannot reveal anything about Japan as a whole. Maybe I should not try to derive any conclusions from it beyond the group studied. Those who want to maintain the centrality of Japanese culture as the primary explanatory variable in studies of the Japanese legal system might find such conclusions convincing. They might say that I was only talking to misfits and if I had talked with others I would have found out how marginal the people I studied are. Or they might say that one does not discover the operation of cultural factors by looking only at "trouble" situations because culture may be most powerfully expressed in its least visible form by the ordinary decisions that people make when there is no trouble.

Another possibility is that the people I studied have as much cultural capital as anyone else in Japan, but that their circumstances were extreme, leading to uncharacteristic actions. Most of the people in my study were from working-class backgrounds and lived in working-class neighborhoods. Their manners seemed to me somewhat more down to earth than those I observed in more elite locations and circumstances. Nevertheless, the politeness, the repeated expressions of apology that pepper Japanese conversation, the characteristic bowing that accompanies greetings, expressions of thanks and departure: All of these gestures were typical of the litigants I studied. So the notion that litigants are a deviant class within Japanese society strikes me as implausible.

But what about the notion that the litigation was a response to extreme circumstances? Statistically, we know that litigation is infrequent in Japan (if we assume that American litigation rates are "normal").[20] Does that mean that any instance of litigation only occurs when people are driven to desperation? This line of argument can quickly become tautological. It is also unhelpful in understanding the processes involved in reaching the point of filing lawsuits, or in continuing the fight throughout the years of delay that are involved. The notions of "extremes" and "desperation" are also difficult to find in the voices that I heard in my research. The sense of outrage at the polluters and the government is muted in the legal documents filed, as it was in the interviews I conducted and my observations. The tragedy of victimization is always present in conversation and action, but litigation in Japan is a slow process of strategic maneuvering, not a radical act designed to get instant results.

Tentative Answers

So far I have only presented some problems of ethnographic research in the attempt to study societies that proclaim themselves to be law-avoiding. The tentative answers

I will discuss here may be relevant not only in the case of work being done in the law and society field, but in sociological research in general.

As to the question of whether I negate the "voice" of the Amish by sticking with the sociolegal framework which I brought to my research, I think this is just a subset of a problem anyone doing ethnographic research faces. If you are doing ethnography, it is, presumably, because you intend to present what you learn about one group to other people so that they can better understand both the group in question, and some aspect of social process in general. That is what social scientists do and it always involves "imposing their view" on the rich variety of observations that they make. Compared with a random sample survey of thousands, which has been criticized for imposing the researcher's voice on respondents, sociological field methods may be less constraining on the ability of people to speak to us in their own "voices." But when you read any kind of ethnographic work, I think it is naive to think that you are hearing someone's "true voice" as if undistorted by its filter—that is to say, the ethnographer.

I find it helpful here to consider Wolcott's (1995) emphasis on the artistic, as opposed to the scientific, quality of ethnographic work. By calling sociological fieldwork an "art," Wolcott is contrasting it with the attempt of quantitative sociologists to locate sociology in the "sciences." In doing so, he relies on a view of art and science as polarized. However, I would substitute the word "creative" for "artistic," and "routinized" for "scientific" in his description because science is done in many ways that range from the drudgery of repeated experiments following strict protocols to the sudden, and definitely creative "aha" experience of recognizing a previously unknown pattern. Anyone who reads my work on the Amish and/or the Japanese will have a unique response to it, depending on his or her prior study of social science issues, law and society debates, religious sects such as the Amish, and the rich and growing body of literature on Japan. Is this uniqueness of response different from what we expect of people encountering a work of art or music or dance for the first time? I agree with Wolcott that there is not a very big difference. The epistemological premise behind the routinized rules of science, which distinguishes this side of science from art, is that anyone following the same procedures used by one scientist will achieve the same results as that scientist has because the method is objective, the real world is stable and knowable, and it does not matter who is looking.[21]

Ethnographic methods in Sociology, on the other hand, require the ethnographer to execute a complex and personally engaging process of developing what Weber called "*verstehen*"—an ability to see the world the way one's subjects see it (1958, orig. 1918). There is both an objective and a subjective component to this process, and trying to untangle these two aspects is similar to an art critic trying to parse what is "accurate" in a painting from that which represents the artist's "interpretation" or "expression." Since in sociology ethnographers compete more directly with proponents of "pure" science (what I am calling the routinized side of science) than with admirers of "pure" art (the idea of art as a purely individualized, subjective, creative process that cannot be reduced to formulas), I suggest that ethnography lies somewhere between the two extremes, being neither pure art nor pure science. As in studying a work of art, one's understanding of an ethnographic report is enhanced by knowing a great deal about the ethnographer.[22] As with art, it is also possible to have a "personal" response to an ethnography without prior knowledge about the ethnographer.

From the standpoint of methodology in the social sciences, the conditions I have described here place a heavy burden on the ethnographer as a person the reader must trust, both in terms of observation and interpretation. Few rules tell the ethnographer how to interpret a statement or a particular action. The individual observations must be repeatedly interpreted within the context of all observations and the ethnographer labors to create a plausible interpretation of them that does not contradict any of the observations.

For the ethnographer's reader, facts do not present themselves regardless of the author's identity and purpose. Ethnographies do not produce tables that allow of only one interpretation. Therefore, it is imperative that the ethnographic report contain as much detail as possible regarding how the field work was done, what the field worker's position in the site was, and what kinds of interpretation are being given to specific, clearly identified pieces of evidence. In addition, it needs to present personal information about the author to the extent that such information may bear on the development of the project and interpretations of observations.

There is another important factor, one that adds value to the ethnographer's work by addressing the problem of "voice." The sociologist Alfred Schutz introduced what he called the sociologist's position as "stranger" (1978; Schutz and Gurwitsch 1989). He was speaking about the semi-alienated position of the sociologist in her or his own society, the simultaneous involvement and distance that draws people into sociology and makes them capable of seeing things as noteworthy that others take for granted or choose not to talk about. I think the same can be said of anyone doing comparative ethnographic research. Just as sociologists as "strangers" are marginal in their own societies, just as they listen to the "voice" of the people they study as if those voices were apart from them, the comparative ethnographic process gains value because the sociologist does not simply report what people say and do. The sociological advantage comes directly from *being* a "stranger," from perceiving what people take for granted as necessary to understanding what they do, but at the same time approaching research situations with a sociological slant and neither privileging nor dis-privileging the world view of the people among whom the sociologist is conducting research.

With respect to the problems I have outlined in the study of Japanese litigation, I come to the same conclusion, though with an added query. The proponents of cultural explanations for something as specific as the condition of the legal system in Japan set culture up as an independent variable in a scientific model of investigation. I would ask: What happens to the credibility of that model if, the closer we look, the more its elements seem to disappear? I have observed Japanese people who, when trouble arises, pursue legal confrontation, disucss it in terms of legal strategy, and consider avoiding it only if it should prove too expensive in time and/or money terms. I have heard them speak of their rights being violated, and express their pursuit of justice. Either their views must be characterized as not Japanese (a circumstance that certainly would call for explanation), or their "Japaneseness" is not relevant to the behavior in question.

Therefore, I wonder whether culture's disappearing act is unique to the issue of law's relationship to culture, or whether this paradox is a more general problem of theory and method in the social sciences. What are we to make of the possibility that, using ethnographic methods that seem to have been tailor-made for the comparison of cultures and were based on the logic that individual bits of data had to be seen within the context of an entire world view, we arrive at a conclusion that makes culture an unnecessary concept in understanding what we are seeing? If we understand enough about a group's situation and their view of it, and their behavior makes sense to us as a rational response to opportunities and restraints perceived by the people we observe, have we simply "gone native"? Are we, as social scientists, capable of adding anything to the account that isn't already there in the accounts we hear? What happens to our position as "stranger" when we eliminate culture as the explanatory variable?

Appendix

Personal Background and Involvement in the Research

In the interests of practicing what I was preaching above, the following section spells out some relevant information about the way I conducted the research I described in this chapter. It also contains information about me that strikes me as possibly having had an impact on the way in which I worked and the ways in which people reacted to me.

My involvement in research on the Amish was the result of contact that developed between Dr. John Hostetler and me over a period of several years. Hostetler, who had a joint appointment in Temple University's departments of Anthropology and Sociology, has written extensively about communal societies in North America and is best known for his work concerning the Amish. He occupies a unique vantage point for applying social science thought to the study of Amish communities. He was born and raised in an Amish family. He, however, voluntarily chose to pursue higher education rather than joining the Amish church. In addition, he has developed and maintained extensive contacts with Amish families and communities throughout the United States and Canada, and continues to be an advocate for Amish groups in their dealings with the outside world.

I had known John Hostetler for years prior to collaborating with him on the research reported here. For one thing, my family and I belong to an "intentional community," a kind of commune. Hostetler has spent his professional career studying communities of all sorts, so he was interested in hearing about my community when he found out I lived there.

In addition, I was coteaching a course on "Comparative Dispute Settlement" at Temple's Law School with a law professor (Peter Severeid) and an anthropologist (Professor Peter Rigby). As part of the course, we invited Dr. Hostetler to guest lecture on "dispute settlement" in Amish communities. These lectures led me to begin discussing a joint project with him, as he had never conducted research specifically on the question of how the Amish deal with conflict, although he had plenty of anecdotal information. These discussions led to a research proposal that the National Science Foundation funded.

The field work lasted for about two years. It consisted of a variety of activities, many of which involved both of us. Most of the activities took place in Lancaster County, which is about seventy miles from where both Hostetler and I lived at the time. Because of the distance, most of our observations took place during day-long trips to the area. On some occasions, however, we stayed overnight in a local Mennonite-owned lodge.

Dr. Hostetler facilitated my entrée into this field with introductions. These introductions were essential because I was (and still am) "English," both in the sense that I was not Amish and did not grow up in an Amish family, and in the sense that I could not speak the Amish version of German. The Amish have a reputation for being polite but distant in dealing with the "English." Many prefer to avoid contact as much as possible. Hostetler advised me on ways that I could make it easier to be accepted. The most important fact would be that he would be introducing me to people who knew and trusted him. In addition, he suggested that I wear some kind of hat during our visits, because Amish people feel that men should wear hats. He also regularly introduced me as coming from a Quaker farm " ... over in Bucks County." My Quaker identity was important because many of the Amish men and women we met were keenly interested in their own history and part of that history intersected in important ways with Quaker groups who have shared the Amish view that war is never justified. It was also important because it connected me to some kind of church, and so I did not come across as uncommitted to anything familiar to them. By introducing me as Quaker, Hostetler was making me both more interesting (because I could tell things about what the Quakers in Bucks County were doing) and somewhat distinguishable from the rest of the "English."

I also learned from observing Hostetler's way of interacting with Amish people. He pointed out how important it was to engage in small talk about family. Moreover, it was obvious from Hostetler's own actions that it was best to avoid diving directly into the issue we had come to discuss. I learned a great deal of patience in the art of indirection in these encounters. Our research was based on field work involving a variety of contacts and visits. Prior to some of these visits, we identified issues that we wanted to hear people talk about. Because of his long-term involvement with the Amish, his position as a sympathetic outsider, and his status as a college professor, Hostetler was at times called on by Amish people for a variety of favors, some of which were rather unique. For example, he could recommend hospitals or medical services, including psychiatric help, to families in need. If someone needed a ride, Hostetler would act as chauffeur, using his own car (this is not one of the unique services, since Amish regularly rely on non-Amish friends for this help, but I mention it to illustrate the kind of contact he has with Amish people). In addition, when Amish interests came into conflict with government agencies or regulations of one sort or another, Hostetler would often help "run interference" through some of the many contacts he maintained in a variety of government offices.

One kind of service he performed was what we would call mediation. On occasion, Amish families or groups invited him to listen to problematic situations, talk with those involved, and suggest ways for them to work out the problem. Some of these were church-wide issues, while others involved tensions within extended families. As a result, he was well-informed about a wide range of problems, both old and new, that were directly relevant to the purpose of our research. Many of the problems and issues we pursued had originally come to his attention in the normal course of his contacts with Amish families. Others came up unexpectedly during our field work.

Some field work occasions took place in barns or on front porches, and were largely just rambling conversations that sometimes included our interjected questions about particular cases (problems that were being circulated in Amish conversations) or about situations that had people worried. These were unstructured conversations, but we knew what issues we wanted to learn about. We had similar conversations in less traditional settings—Amish factories (e.g., a farm implements manufacturer, a garden furniture workshop, and a buggy-builder's workshop), small restaurants, and a hub-cap shop being run by the non-Amish son of an Amish farmer.

In other cases, we were participating in social events that involved people Hostetler thought I should get to know. We did not attend these in search of specific information about trouble, though we often heard about problems anyway. We had dinner in one home, for example, with about twenty people from three families and I happened to be sitting next to a young woman who had had the disturbing experience of discovering an Amish murder victim. Our early polite conversation led inadvertently to this subject (I had not even heard of the event and had not expected to hear such a thing) and she spent about forty-five minutes describing this experience in detail.

We also traveled to Harrisburg (the state capital) in order to interview government officials who had had direct dealings with Amish representatives. These were open-ended interviews that we conducted together.

On some occasions, I spent time alone with Amish groups or individuals. For instance, I spent a whole day traveling with a group that came to Philadelphia to see some sights. We gathered originally in a Mennonite home in Philadelphia, and I went with them as they visited a park, a seventeenth century Mennonite church, a historically interesting moored sailing vessel, a farmer's market where some Amish acquaintances were selling farm produce, and other historically significant places in the city. On such occasions, I would share my observations with Hostetler and we would discuss their significance for the research.

The field work was nearing its conclusion when I was called on to move to Japan to teach in Temple's program there. I therefore had to leave the field quite abruptly. Before doing so, however, we made several more trips to Lancaster County. I also returned with Hostetler during the following summer while I was back in the United States for vacation. After four years in Japan, I returned to Philadelphia, and have made additional visits to the area, in one case to participate in a scholarly conference on the Amish, in which a number of local Amish people participated.

Our analysis of the field notes and the writing of articles for publication took place through long-distance consultation between Hostetler and me. In addition, Hostetler came to Japan during a visit he was making to a Hutterite community outside of Tokyo, and we worked on the research while he was there.

Research in Japan

My reason for being in Japan was, as I explained above, to teach in Temple University's undergraduate program there. I had taught in the program for one year at the program's inception (1983–1984). During that year, I studied Japanese and practiced whenever I could in encounters with shopkeepers and others, but because the Temple program required faculty to teach in English and work in an English-speaking environment all day, my ability in Japanese remained limited. After returning to Philadelphia, I continued to take Japanese lessons for the following four years and was able to complete Temple's most advanced regular courses prior to my return in 1988. I continued taking lessons during this second stint, but once again I was working in an English-only environment and this fact restricted my language learning. The level of my commitment to language learning was also limited by the fact that each year in the program was always defined as my last year there, until I could extend it late in the year. As a result, I was always of a mind that this might be my last few months in Japan and I might never have the opportunity to come back.

I have described this experience in detail because language is an issue that many scholars who do research in Japan consider important, if not downright essential. I did not get into doing research in Japan in the traditional way, studying the language thoroughly before going, immersing myself in a Japanese-only environment, and focusing my attention on traditional issues of interest to scholars of Japan. I estimate that by the time I began my research there in early 1989, I had achieved about a fourth-grader's level of Japanese. I continued to learn Japanese both formally and informally during the research, but I cannot say that I ever reached a level of fluency in the language that would allow me to understand subtle nuances of meaning. My Japanese was not good enough to carry on open-ended interviews without assistance, much less participant observation. This fact puts limits on the research I was able to do, and it required numerous adjustments, as detailed below.

The research that I describe in this article began as a result of conversations I had with Professor Setsuo Miyazawa of the Law Faculty at Kobe University. I had gotten to know Professor Miyazawa through his participation in the Law & Society Association. Even before I arrived in Tokyo in 1988, he and I talked about doing some collaborative research. He contacted me with "hot" information about a lawsuit that had just been filed in

Kobe District Court by a group of air pollution victims, and suggested that this might be a good case to study.

So, as in the case of my research on the Amish, I became involved in the air pollution litigation research through contact with an "inside" scholar. In his position as both a prestigious and active member of the Kobe law community, he had already developed contacts with many of the lawyers who were involved in the lawsuit. Miyazawa arranged several early meetings with leaders of the litigants' group and with their lawyers. Because Kobe was 330 miles away from my home in Tokyo, these early meetings had to be carefully scheduled. There were only limited opportunities to "hang around" and engage in casual conversations.

In addition, because of my limited Japanese ability, Miyazawa and I attended all these early meetings and he acted as translator for me. My role in these meetings was to ask the questions, and his role was to explain my questions to the interviewees and tell me their answers. Knowing the subject of the questions, I could often understand much of the answers being given, but not enough to enable me to do the interviews alone. We taped these interviews, and went over the tapes afterwards to fill in information that we might have missed in the original translation. I transcribed all of these tapes, adding in my own observations of nonverbal cues such as body language, physical surroundings, smells, noise, etc. We found that the disadvantages of my linguistic limitations were at least somewhat offset by the advantage we gained through this translation process. While Miyazawa was carrying out the translation, I had time to assess what we had already heard and develop questions that reflected the answers to previous questions. In addition, he could follow my line of questions and suggest additional questions that I "might want to ask" based on his own knowledge of the subject. We both agreed that these interviews ended up being stronger than they could have been if either of us were doing them alone.

The question of entrée is more complex in this case than it was in the Amish research. Being American, I was obviously an outsider, as I was in Lancaster County. Even with fluent Japanese I would have been seen that way. Because of Japan's long and sometimes stormy relationship with the United States, many Japanese, like many Amish people, have come to rely on a standard repertoire of ways to be polite yet protective of their own privacy when dealing with Americans. On the other hand, Professor Miyazawa found that because I was an outsider, it was easier for him to get some kinds of access for these interviews than it would have been if he had tried to do it alone. His social position as a Japanese academic meant that he faced barriers of protocol and, perhaps, trust that my outsider status lacked. It seemed to make sense that a foreigner might want to learn things about Japan, and that a foreigner would not be likely to have the kinds of social ties and obligations that would complicate matters in talking to a Japanese professor. Hence being an outsider had its advantages.

Nevertheless, in order to gain any access at all, I had to submit a full description of my research objectives and plans to the head lawyer in the case. He in turn brought it to a meeting of key members of the lawyers' team and leaders of the air pollution victims' association. They had to decide whether to allow any research at all. When they finally contacted me, telling me that they approved of the research, they did so without attaching any restrictive conditions.

On the other hand, the level of entrée that we did achieve came at a cost that is perhaps typical in research on highly visible cases of litigation. We judged that our entrée with the litigants might be compromised if we tried to gain the same kind of access to their opponents in the litigation. I attended court hearings at which as many as sixty litigants made their presence loudly known. I was there as the TV cameras and photojournalists did their work. I sat with plaintiffs' lawyers or in the section reserved for plaintiffs during the hearings. In one hearing I sat beside a litigant who was in pajamas in a wheel chair and breathing oxygen from a portable tank. I was therefore very conspicuous to the defendants' lawyers as being on the plaintiffs' side. The same would have been true if I had come into court sitting with the defendants' lawyers. As a result, all of our research centered on the activities of the plaintiffs' side. This research produced interesting conclusions that could be more clearly understood if we could find out how the defendants' viewed the situation, but we sacrificed this perspective in order to maintain entrée with the litigants. For general background, I interviewed the heads of the legal department of a large corporation that occasionally finds itself involved as defendant in cases such as those I was studying. But that corporation was not involved in the cases that were the focus of our study. After a couple of years of interviews, visits to court during hearings on the case, and late-night visits to bars with some of the lawyers involved, Professor Miyazawa and I applied for an NSF grant to continue the research, and the grant was approved. The funds allowed me to make regular visits to Kobe. Our research expanded to include more of the lawyers who were working on the case, doctors and others who had been involved at the outset in identifying the air pollution disease problems, and a large set of litigants. In many of these latter interviews and observations. Miyazawa introduced me to bilingual graduate students who served as translator. These students were from English-speaking countries and were earning law degrees at Kobe University. The procedures for these observations and interviews was usually the same as with Professor Miyazawa—I would ask the questions and the student would translate both ways, with a tape recorder serving as backup.

One set of observations and interviews came from an unusual opportunity. We were able to accompany lawyers on their visits to litigants' homes to go over the "stories" the litigants would tell in court testimony. These visits usually lasted about two hours. The lawyers (usually two, a senior and a junior) went into great detail about how the litigant should word descriptions of their conditions and problems during testimony. They also discussed what litigants should avoid revealing if possible (e.g., the fact that the litigant was a chain smoker who could not stop even though ordered to do so by the lawyers). The lawyers were using these sessions to make sure that all sources of testimony agreed with each other, and to decide which of the 400-plus litigants to put on the stand when the testimony phase began.

In each of these visits, after the lawyers were finished with their questions, they would turn to me and ask if I had any questions. These visits therefore constituted an unusual opportunity for observation (lawyer–client interactions, narratives being developed for the courts, interactions among the family members present, conditions under which litigants were living, etc.) and interviews. The lawyers found my questioning sometimes valuable for themselves because they learned things about the way their clients felt that they had not been able to discover with their own questioning. Traveling to and from the interviews with the lawyers also provided extended opportunities to find out how the individual lawyers viewed their own roles in the process, how they fit this work in with the rest of their practice, what they thought of the people they were representing, and other aspects of their definitions of the situation.

It is fair to say that I developed a great deal of sympathy for these litigants and respect for their lawyers. On one occasion I accompanied the president of the litigants' group to the top of a pedestrian overpass so that she could show me the traffic on the sixteen-lane highway below and the smokestacks of the nearby defendants' factories. She insisted on doing this despite struggling with asthma to climb those stairs. Her rasping breathlessness as she reached the top of the stairs was nearly as eloquent as the sweep of her arm as she gestured at the quivering, exhaust-filled air that hovered over the area. She was determined, despite her asthma, that I should see the situation myself.

In litigants' homes, I saw the mundane struggles of the litigants and their families in dealing with the limitations imposed by their diseases. I was impressed by the stoicism and determination most of them displayed in spite of these obvious problems. One of these was literally a deathbed interview—the elderly litigant lay on his futon throughout the interview and was constantly coughing up foul-looking phlegm. His adult daughter attended to his needs throughout the interview. Nevertheless, he was determined to be heard. Two days after the interview he died.

It is also fair to say that one reason for the openness with which the litigants' group welcomed me, aside from Professor Miyazawa's intervention, is that I was perceived as a potential ally in publicizing their plight and spreading the word about the justice of their cause.

As I concluded in the research, an important part of litigant strategy in lawsuits such as these is the application of constant pressure on the central and local governments to change policies to protect working-class people. Because the courts themselves are so slow and restrictive in the remedies they offer, such people use litigation as a form of political action that pressures governments to alter policy and/or implement promises. International perspectives influence these actions. Lawyers and litigants invoke international environmental standards and legal ideas as part of the pressure they try to exert on local governments. For some in these groups, the objective is even broader—to build international coalitions of activists who can work on environmental problems at the global level. Leaders of the litigants plied me with sophisticated publicity material, including glossy English-language booklets, and professionally produced videotapes dealing just with their lawsuits. One lawyer even told me that she had seen one of my publications about the case and was glad to talk with me because, on the whole, I had presented the litigants in a sympathetic light. These actions fit in with the glossy color photographs and videotapes as part of a way to extend their base of support and invoke internationally legitimized ideologies about environmental issues. To some small extent my research may have served these interests, though it would be presumptuous and inaccurate to claim any widespread impact, considering how small the audience is for most sociological research.

Notes

* Chapter originally presented at the June 1998 Annual Meetings of the Law and Society Association in Aspen, Colorado.
1. I enclose this word in quotation marks because I do not mean for it to denote the full range of activities of some early Anthropologists (e.g., Murdock 1960) who intended to record the details of precapitalist cultures before they were obliterated by advancing capitalism. Nor am I referring to the full range of activities other anthropologists have included under that term, including observation, interviewing, and archival research. Rather, I use it in the looser sense in which sociologists have incorporated the word to add to their vocabulary as a means of distinguishing their work from quantitative hypothesis testing. When sociologists do what they call ethnographic research, they often mean observational research, perhaps accompanied by open-ended interviews that stem from the observation. They usually go into such research without systematic pre-existing hypotheses, though they often admit to having "hunches" about what they are seeking. As I mention in this chapter some consider this activity science, while others liken it more closely to an artistic or literary activity.
2. Unless otherwise noted, when I speak of the Amish, I am referring to what is generally known as the "Old Order" Amish, in distinction from less conservative groups.
3. See Ewick and Silbey (1998: ch. 3) for a comprehensive review and discussion of the many ways the term "legal consciousness" has been used.
4. See Nelken (1997) for a current sampling of the debates on the meaning and value of the concept "legal culture." See especially his introduction for a summary and classification of the different approaches.
5. Felstiner's (1974) term, as in "like it or lump it."
6. These details on Amish life are derived from the works of John Hostetler (1993) and Donald Kraybill (1989). In addition, my pre-research expectations about Amish society were formed by extensive conversations with both Hostetler and Kraybill. For a fuller description of this ideology, see Kidder and Hostetler (1990).
7. Structural-functionalism refers to the line of thought developed by Durkheim (1964, orig. 1895; 1964, orig. 1893), Parsons (1954, 1964, orig. 1951), and Merton (1968).

8. Conflict theories share their origins from the works of Marx and Engels (e.g., 1967, orig. 1867), and W. E. B. Dubois (1940, 1967, orig. 1899), and center on the role of power (e.g., see Mills 1956).

9. See, for example, Becker (1951, 1953, 1963, 1982), Becker et al. (1961), Liebow (1993, ch. 1) for specific examples of research conducted on the basis of this theoretical position. See Mead (1962, orig. 1934) for an early statement of the theory.

10. See Ragin and Becker (1992) for a discussion of the epistemological implications of "cases."

11. By "sincere" I do not mean a response that is more believable or true. I am referring rather to an interpretation that the response reflects no intervening instrumental calculus about what may be gained or lost in saying what is said. Interpreting a response as "sincere" simply means taking it at face value as reflecting the respondents overall view of a situation.

12. Wisconsin v. Yoder, 406 U.S. 205 (1972) Note that this case was not brought by the Amish themselves, but by sympathetic "friends" who provided all the financial and professional assistance needed. The case resulted from the refusal of an Amish farmer to follow the state's mandatory schooling requirements.

13. The Amish farmers argued that the manure management requirements would be so costly that they would drive more Amish men off farms. The Amish believe in general that their ability to adhere their faith depends strongly on their ability to continue farming as their principle economic activity.

14. John Hostetler was instrumental in helping me gain entree into the Amish communities of Lancaster County, Pennsylvania and participated in many of the observations and interviews that made the basis for our collaborative research.

15. For details of this research on a "class-action" type lawsuit filed by 500 members of an Air Pollution Victims Association, see Kidder, R. and Setsuo Miyazawa (1993).

16. Culture was something shared by both those who litigated and those who did not. There was no reason to think that the litigators were any less "Japanese" than their nonlitigating neighbors. What I found was that structural barriers to litigation, such as its high cost and the unavailability of lawyers (produced by rigid controls on the size of the legal profession), held more promise than culture as a means of explaining why people chose not to litigate.

17. My informants filed their suit in 1988. In February, 1999, they reached a partial settlement with the nine private companies, but have not yet settled all issues with the government entities.

18. These findings would reinforce the point that John Owen Haley (1978) originally made in his argument against culture-based theories of Japanese legal development.

19. These surveys involved hypothetical situations that raised questions about what the interviewees would consider "just" outcomes. The research was designed to compare concepts of justice in Japan and the United States.

20. The comparative infrequency of Japanese litigation is by no so taken-for-granted among both law and society scholars and those who specialize in Japanese studies that the only question is how it should be explained. See Haley, 1978 and 1991 (ch. 5); Ramseyer (1988).

21. Wolcott's position, and my modification of it, are both informed by the critiques of "pure" science initiated by Kuhn (1962, 1970).

22. It is this important additional information about the author, for example, that was recently (1998) praised by the Law & Society Association when it awarded Barbara Yngvesson a prize for her article (1997) about "open" adoptions. Our understanding of her research and her interpretations is more complete when we understand her position as an adoptive parent.

References

Becker, Howard S.
   1951   "The Professional Dance Musician and His Audience." *American Journal of Sociology* 57: 136–144.

1953   "Becoming a Marihuana User." *American Journal of Sociology* 59: 235–243.
1963   Outsiders: Studies in the Sociology of Deviance. New York: Free Press.
1982   Art Worlds. Berkeley: University of California Press.
Becker, Howard S. et al.
1961   Boys in White: Student Culture in Medical School. Chicago: University of Chicago Press.
Engel, David M.
1984   "The Oven-Bird's Song: Insiders, Outsiders, and Personal Injuries in an American Community." *Law & Society Review,* 18(4): 549–579.
Ewick, Patricia and Silbey, Susan
1998   The Common Place of Law: Stories from Everyday Life. Chicago: University of Chicago Press.
Greenhouse, Carol
1986   Praying for Justice: Faith, Order and Community in an American Town. Ithaca: Cornell University Press.
Felstiner, William
1974   "Influences of Social Organization on Dispute Processing," *Law and Society Review,* 9: 63.
Goffman, Erving
1959   The Presentation of Self in Everyday Life. Garden City, NY: Doubleday.
Haley, John O.
1978   "The Myth of the Reluctant Litigant and the Role of the Judiciary in Japan." *Journal of Japanese Studies* 4: 359–390.
1991   Authority Without Power: Law and the Japanese Paradox. New York: Oxford University Press.
Hamilton, V. Lee and Sanders, Joseph
1992   Everyday Justice: Responsibility and the Individual in Japan and the United States. New Haven, CT: Yale University Press.
Hostetler, John
1993   Amish Society (4th ed.). Baltimore, MD: Johns Hopkins University Press,
Kawashima, Takeyshi
1963   "Dispute Resolution in Contemporary Japan." In A. T. von Mehren (ed.), Law in Japan: The Legal Order of a Changing Society. Cambridge: Harvard University Press.
Kidder, Robert
1973   "Courts and Conflict in an Indian City: A Study in Legal Impact." *Journal of Commonwealth Political Studies,* 11 (2). Reprinted 1987 in Yash Ghai, Robin Luckham, and Francis Snyder (eds.), The Political Economy of Law: A Third World Reader. Delhi: Oxford University Press.
Kidder, Robert and John Hostetler
1990   "Managing Ideologies: Harmony as Ideology in Amish and Japanese Societies." *Law and Society Review* 24: 895–922.
Kidder, Robert and Setsuo Miyazawa
1993   "Long-Term Strategies in Japanese Environmental Litigation," *Law and Social Inquiry,* 18(4): 605–627.
Kraybill, Donald B.
1989   The Riddle of Amish Culture. Baltimore, MD: Johns Hopkins University Press.
Kuhn, Thomas S.
1962, 1970   The Structure of Scientific Revolutions (2nd edition). Chicago: University of Chicago Press.
Liebow, Elliot
1993   Tell Them Who I Am: The Lives of Homeless Women. New York: Free Press.

Nader, Laura
1990   Harmony ideology: Justice and Control in a Zapotec Mountain Village. Stanford, CA: Stanford University Press.

Mead, George Herbert
1962, orig. 1934   Mind, Self & Society from the Standpoint of a Social Behaviorist, Charles W. Morris (ed.) Chicago: University of Chicago Press.

Murdock, George Peter
1960   Ethnographic Bibliography of North America. New Haven, CT: Human Relations Area Files.

Ragin, Charles C. and Howard S. Becker
1992   What is a Case? Exploring the Foundations of Social Inquiry. Cambridge: Cambridge University Press.

Ramseyer, Mark J.
1988   "Reluctant Litigant Revisited: Rationality and Disputes in Japan." *Journal of Japanese Studies* 14(1).

Schutz, Alfred
1978   The Theory of Social Action: The Correspondence of Alfred Schutz. Bloomington: Indiana University Press.

Schutz, Alfred and Aron Gurwitsch.
1989   *Philosophers in Exile: The Correspondence of Alfred Schutz and Aron Gurwitsch, 1939–1959,* Richard Grathoff, ed.; J. Claude Evans, trans.; foreword by Maurice Natanson. Bloomington: Indiana University Press.

Taylor, Jill M., Carol Gilligan, and Amy M. Sullivan
1995   Between Voice and Silence: Women and Girls, Race and Relationship. Cambridge, MA: Harvard University Press.

Wagatsuma, Hiroshi and Arthur Rosett
1986   "The Implications of Apology: Law and Culture in Japan and the United States," *Law & Society Review* 20(4): 461–498.

Weber, Max
1958, orig. 1918   "Science as a Vocation" In H. H. Gerth and C. Wright Mills, From Max Weber: Essays in Sociology. New York: Oxford University Press, pp. 129–156.

Wolcott, Harry F.
1995   The Art of Fieldwork. Walnut Creek, CA: AltaMira Press.

Yngvesson, Barbara
1997   "Negotiating Motherhood: Identity and Difference in 'Open' Adoption" *Law & Society Review* 31(1): 31–80.

CHAPTER 6
RECONCEPTUALIZING RESEARCH:
ETHNOGRAPHIC FIELDWORK AND IMMIGRATION
POLITICS IN SOUTHERN CALIFORNIA

Susan Bibler Coutin

With the publication of *Argonauts of the Western Pacific* in 1922, Bronislaw Malinowski set the standards for ethnographic research for decades to come. His advice to ethnographers was straightforward. Know the natives. Live among them. Adopt their point of view. Learn their language. Spend a long time in the field. Take copious fieldnotes. Locate and interview key informants. Produce data-driven accounts. While the authoritativeness of ethnographic texts has recently been questioned[1] and while most anthropologists no longer regard their research subjects as "natives," these fieldwork practices remain central to ethnographic research on legal and other topics.

The purpose of this chapter is to consider the ways that such enduring ethnographic practices are redefined by the research context and analytical tradition in which they are located. In addition to being a research technique, participant observation places an ethnographer among a group of people who are engaged in a particular set of activities. An ethnographer's presence may have serious political and legal consequences for the ethnographer and others (Coutin and Hirsch 1998; Nordstrom and Robben 1991). Similarly, conducting an interview is not only a means of generating "data." In addition, such conversations have antecedents, create records, and enter ongoing debates. The significance of "research activities" is determined not only by ethnographers, but also by research participants, government authorities, political activists, attorneys, and others. Ethnography cannot be isolated from the processes that make ethnographers perceive particular topics as worthy of attention

To examine how particular research contexts redefine ethnographic fieldwork, I discuss the ways that immigration politics in southern California shaped my 1995–1997 fieldwork regarding Central American immigrants' efforts to obtain legal status in the United States. Through this study, I sought to identify ways that unauthorized immigrants—who have often been deemed relatively powerless players within economic and political processes—negotiate U.S. immigration categories and policies. My goals were to learn how immigrants were affected by their legal statuses, identify the legal and political strategies devised by the undocumented and their advocates, and assess the efficacy of these strategies. To these ends, I employed the ethnographic techniques of interviews and participant observation. Specifically, I volunteered with the legal services programs of three Central American community organizations in Los Angeles, observed deportation hearings in an immigration court, participated in community-wide events and demonstrations, and interviewed immigrants, activists, and attorneys. These research activities were given other meanings by the context in which they occurred—and by my position within that context. For example, conducting an interview in Spanish, attending a demonstration, and helping an immigrant apply for political asylum were (not surprisingly) political acts. Similarly, observing a court hearing, recording information about an immigrant's legal case, and filling out a work permit renewal request were legal activities. In fact, through the use of signed consent forms, written documentation, and the creation of "records," law and ethnography employ similar understandings of "truth" and "authorization" (see Coombe 1998). To view fieldwork as merely or even primarily research can underestimate the complexity and embeddedness of social interaction and overestimate the power of ethnographers.

My strategy in this chapter is to use jarring or uncomfortable fieldwork moments to expose the multiple meanings of participant observation and interviews. I begin by describing the political context in which I did research. This context was characterized by increasing polarization around immigration issues. I then briefly discuss recent analytical moves that shaped my understanding of "identities" and my approach to fieldwork. The bulk of this chapter is devoted to an analysis of the ways that immigration and identity politics redefined interviews and fieldwork. I focus particularly on research subjects' reasons for agreeing to participate, the politics of language, race, and class, and the legal and political significance of my presence as an ethnographer. In more standard methodological treatises, these topics are often described as the need to gain access, establish rapport, and efface presence. In my own experience, the methodological dimensions of such research moments often faded in comparison to research subjects' desire to counter anti-immigrant stereotypes, the salience of racialized and other discourses, and the legal ramifications of recording particular statements or actions. Clearly (and again, not surprisingly), the meanings of "research" are inextricable from the politics of knowledge.

Immigration Politics in Southern California

By 1995, when I began fieldwork within Los Angeles-based Central American community organizations, immigration politics were extremely polarized (Perea 1997). In the early 1990s, unemployment, economic restructuring, and demographic changes fueled anti-immigrant sentiment and led to calls for more restrictive immigration policies. In 1993 California voters passed Proposition 187, which required teachers, health care workers, and other social service providers to verify the legal status of their students, patients, and clients, and to report suspected illegal aliens to the U.S. Immigration and Naturalization Service. The constitutionality of Proposition 187 was immediately challenged, which led to a court injunction against enforcing most of the proposition's provisions (Martin 1995). Controversy over immigration policies continued, with heavy news coverage of immigration issues. Studies debated whether immigrants were using welfare, assimilating, paying taxes, taking away citizens' jobs, and abusing public services. Politicians added immigration reform to campaign promises. In private conversations, I found that many friends, acquaintances, students, and relatives were outraged by what they perceived as high immigration levels, and feared that the state, if not the nation, was in crisis. In 1996 legislators responded to such concerns by passing the Antiterrorism and Effective Death Penalty Act (AEDPA), the Illegal Immigration Reform and Immigrant Responsibility Act (IIRIRA), and the Welfare Reform Law.[2] This legislation made legal immigrants ineligible for many public benefits, expanded the range of crimes for which aliens could be deported, and made it more difficult for undocumented immigrants to legalize.

These immigration debates were linked to complicated identity politics. During the 1960s and 1970s national movements that were organized on the basis of gender, sexual orientation, race, ethnicity, nationality and so forth arose to celebrate common histories, confront common oppressions, rectify past injustices and claim civil rights. The 1980s saw a backlash against such movements, with accusations that minority groups were practicing reverse racism and requesting special treatment, calls for "color-blind" procedures and the abolition of "quotas," and the launching of "English-only" initiatives.[3] By the 1990s, with the passage of California Proposition 209, which prevented public institutions from considering individuals' gender, race or ethnicity, it had become possible to couch attacks on affirmative action in the language of civil rights. This move effectively turned the discourse of identity politics in the 1960s and 1970s inside out. In Los Angeles, police officers beat Rodney King, a jury acquitted the accused officers, and the city erupted in race riots. The O. J. Simpson trial added salt to these wounds, with the "race card" being played, celebrated, and denounced. While these issues did not explicitly target immigrants, per se, many immigrants and immigrant advocates experienced racial and ethnic polarization[4] as part of a multifaceted attack on their rights. *La Opinión,* the leading Spanish-language newspaper in Los Angeles, complained that Latinos and Asians were being ignored in the English media's treatment of the O. J. Simpson case as a Black versus White issue. The activists I met during fieldwork attributed restrictive immigration policy to racism. There were blurred boundaries between opposing immigration and opposing affirmative action, multilingualism, or diversity (Heyman 1998).

Immigration and identity politics came together in the legal cases of the Salvadorans and Guatemalans among whom I did fieldwork. Most of these immigrants entered the United States during the 1980s, when civil wars raged in their homelands. At the time, U.S. foreign and refugee policies precluded granting them political asylum, although the earliest migrants were able to qualify for legalization through the 1986 "amnesty" program. By the 1990s legal and political advocacy had resulted in temporary remedies for some 300,000 of these migrants. Salvadorans were granted Temporary Protected Status (TPS) through the 1990 Immigration Act, and then were permitted to register for Deferred Enforced Departure (DED) status. Both Salvadorans and Guatemalans benefited from a legal settlement that gave them the right to (re)apply for political asylum under special rules designed to ensure fair consideration of their claims. Because peace agreements signed in El Salvador in 1992 and in Guatemala in 1996 made it unlikely that many of these immigrants would win political asylum, activists and community members staked their hopes on other means of legalization, such as family visa petitions or suspension of deportation, a remedy available to those who can demonstrate seven years of continuous residency, good moral character, and that deportation would be an extreme hardship.

Immigration reform had a dual effect on Central Americans' pending legalization cases. On the one hand, more restrictive policies made legal status more important, thus increasing the numbers of individuals seeking regularization and delaying hearings and interviews. On the other hand, reform imposed additional obstacles (such as increased fees and residency periods) on legalization applicants. Seeking to be exempted from new, more restrictive immigration policies, Central Americans argued that as a group, they had adapted to the United States, formed strong community ties, and given birth to U.S. citizen children who needed to go to U.S. schools, maintain their English-language skills, and pursue their career goals (Coutin 1998). Similarly, to prove that deportation would be a hardship, individual suspension applicants had to demonstrate that they had acculturated, set down roots, and learned English. To immigrate, these Central Americans had to meet definitions of "Americanness" not unlike those promoted by English-only advocates.

My own interest in analyzing Central Americans' legalization strategies was inspired at least in part by the heightened controversy over immigration issues and the complicated legal situation of Central Americans.[5] To quote the grant proposal through which I secured funding for this project, "examining the identities that Salvadorans are attempting to construct will reveal the ways that Salvadorans are positioning themselves in the United States in the 1990s, and how such positionings engage and counter the increasingly negative popular discourse about immigration in this country" (Coutin 1994: C14). In embarking on this project, I acknowledged that I was sympathetic to immigrants and other subordinated groups, but I distinguished my own research agenda from the public debate over immigration issues. Instead of adopting one or another side within this debate, I hoped to analyze and critique the terms in which the debate was being conducted. I thought that such a project would be useful to those who were seeking to defuse anti-immigrant sentiment. I also assumed that given my prior work on Central American issues (Coutin 1993), Central Americans would perceive me as something of a sympathetic "insider" and ally. These assumptions informed my fieldwork regarding Central Americans' legal identities.

"Identity" as an Analytical Category

I was not alone in perceiving "identities" as a significant analytical and legal category. The identity politics that intersected with immigration policy-making and that affected Central American immigrants' legalization cases had academic counterparts. Since the 1970s interdisciplinary programs and subdisciplinary specializations in feminist theory, critical race theory, ethnic studies, and so forth have proliferated—in many cases through scholars' participation in identity-based movements and organizations (Evans 1979; Matsuda, Lawrence, Delgado and Crenshaw 1993; Rosaldo and Lamphere 1974). The scholars who organized these programs and who developed these specializations use such categories as "women," and "African Americans" to analyze and critique hegemonic structures and to challenge essentializing notions of the self (Yanagisako and Delaney 1995). Many of the contradictions surrounding "identities" and "identity politics" come into focus within law, as the notion that "all are equal before the law" suggests that legal categories are universal, even as law defines the individual as the possessor of "qualities"—including ethnic, gender, sexual, racial, and other "differences"—that must themselves be protected by law (Collier, Maurer and Suarez-Navaz 1995; Macpherson 1962). Claiming rights requires assuming an identity that the legal system can recognize, a process that can simultaneously be dehumanizing and empowering (Coutin and Chock 1995; Merry 1990; Yngvesson 1993). Individuals' consciousness of their right to equal treatment can mobilize them to take legal action, but that action can in turn subject them to the scrutiny of the state that is to protect their rights (Yngvesson 1993). Law is thus one of the arenas in which "identities" of various sorts are denied, assumed, and created.

To analyze the legal "identities" that Central Americans both claimed and were compelled to assume, I drew on insights from social and cultural theory. The most fundamental of these insights is the observation that identities are constructed in a field of power relations (Foucault 1979, 1980; Gordon 1980). The categories through which identities are constructed are, at least in part, products of hegemonic processes such as colonialism and capitalist expansion (Anderson 1991). Assuming particular identities therefore entails being defined according to categories that are integral to discourses of subordination.[6] At the same time, as they challenge the established order, subordinate groups appropriate and redefine what might otherwise be derogatory or restrictive terms. As a result, new formulations of dominant categories are continually produced, and political struggles focus, among other things, on determining whose formulations will be authoritative. Within legal studies, analyses of the definition and transformation of disputes have focused on the the ways that particular legal contests invoke and possibly reshape broader political processes.

My fieldwork was also influenced by recent attention to the spatialization of identities (Anzaldua 1987; Appadurai 1991; Gupta and Ferguson 1997; hooks 1984; Rosaldo 1989). Scholars have noted that spaces, cultures, peoples, and identities have been conflated in both scholarship and popular culture. Such conflations contribute to commonsensical but nonetheless flawed assumptions, such as the notion that "Indian culture" is located almost exclusively in "India" (Gupta and Ferguson 1997) or that refugees who are forced out of their places of birth are "unrooted," suffering from identity confusion, and in need of medical and psychological intervention (Malkki 1992). Within anthropology, spatializations of identity take the form of a "field"/"academy" dichotomy that situates researchers outside of the contexts in which they conduct fieldwork (Coutin and Hirsch 1998; Gupta and Ferguson 1997). In the case of immigration, the conflation of places and persons can either justify exclusionary policies or can enable people who are physically present to claim legal personhood. The identity ascribed to individuals can also redefine the spaces that they occupy, such that, for example, unauthorized immigrants are defined as existing outside of society, in an "underground," or in "the shadows," when in fact, undocumented people live and work among the documented.[7] Through fieldwork I sought to understand how immigration categories shaped social spaces, and vice versa.

My analysis was also informed by theory regarding the limitations and possibility of agency in defining one's own identity. I was therefore particularly interested in immigrants' own efforts to define their legal identities. On the one hand, subjects are invited into being by texts that are already written (White 1985), making subjects inextricable from the processes through which they are produced. As Butler states regarding gendering, "it is unclear that there can be an 'I' or a 'we' who has not been submitted, subjected to gender, where gendering is, among other things, the differentiating relations by which speaking subjects come into being" (1993: 7). Institutions and particularly the state are critical to this process, demanding certain categorizations as a prerequisite for rights and services (Clifford 1988; Engel 1993; Yngvesson 1997). On the other hand, despite institutional constraints, categories of personhood are never completely determined, as critiques of essentialism and of the notion of a unified self have demonstrated (Derrida 1976; Kristeva 1981). Contradictions within particular categories create opportunities for revision and innovation. Being incorporated into an already-written text changes the text over time, even if almost imperceptibly. Subjugated knowledges subvert more dominant discourses and can be sources of innovation. As Anzaldua writes regarding Chicano Spanish: "The switching of 'codes'...from English to Castillian Spanish to the North Mexican dialect to Tex-Mex to a sprinkling of Nahuatl to a mixture of all of these, reflects my language, a new language—the language of the Borderlands. There, at the juncture of cultures, languages cross-pollinate and are revitalized; they die and are born" (1987: iv).

Finally, I took seriously the idea that identities are simultaneously constructed and real. Categories of personhood are material in that they shape individuals' lives in complex ways. If particular designations are prerequisites for claiming rights, then definitions of personhood have material consequences. Categorizations can potentially subject individuals to the power of the state or to being targeted by other groups (Feldman 1991). Certain categories of personhood, such as race or gender, are assumed to be manifested physically. Discourses of power materially shape individuals' actions and their very being, producing the "persons" that disciplines require (Foucault 1979). Racism, sexism, homophobia, and other forms of oppression have very real consequences in the lives of subordinate groups. Individuals also actively participate in the construction and reformulation of their own categorization, making identities real to those whom they define. My fieldwork sought to identify ways that legal categories had shaped immigrants' lives, as well as to the designations that immigrants found most meaningful.

These analytical assumptions led me to focus my research on processes of claiming, disputing, asserting, and even ignoring legal status. I therefore interviewed Salvadoran immigrants regarding their legalization histories, participated in the legal services programs of community organizations that represent Central Americans in deportation proceedings, observed immigration hearings, and followed political campaigns through which immigrant advocates sought to influence U.S. immigration law. As I engaged in fieldwork among Central American immigrants in Los Angeles, I found that my own identity was defined by the processes that I have outlined above. I also discovered that my "research subjects'" understandings of immigration and identity politics sometimes differed from my own. There were moments when these processes and differences defined research as something else.

Interviews and Participant Observation

Although I used interviews and participant observation as research methods, immigration and identity politics sometimes turned my interviews into rebuttals of immigrant bashing and made participant observation a legal and/or political act. The ethical issues that arise in a classic social science research paradigm therefore took different forms in my fieldwork. For instance, methodological discussions often focus on the politics of access, relationship, and presence. First, discussions of access usually concern the need to build trust, accurately represent one's research agenda, and gain research subjects' consent. While these issues were important to me, this representation of "access" fails to consider the possibility that "research subjects" may seek "access" to the researcher for their own purposes. I found that some Central Americans regarded me as a representative of Anglo society through whom they could challenge stereotypical depictions of immigrants. Second, traditional methodological discussions of the relationship between researcher and informant advise the researcher to build "rapport" or a sense of commonality. In my experience, the immigration debate and accompanying social distinctions sometimes overshadowed the interview process. Finally, methodological discussions of ethnographic presence often focus on the paradoxical (and futile) demand that ethnographers simultaneously be present yet absent, observers who participate, unintrusive yet interactive beings.[8] In my own case, a different set of issues arose; specifically, the need to be legally and politically accountable for my presence as an ethnographer. Instead of striving to be, in certain ways, "absent," I confronted the legal and political implications of presence. To demonstrate how other processes and discourses sometimes redefine research, I discuss fieldwork moments in which these issues were exposed.

Access

My research agenda required me to gain access to four overlapping groups: Central American community organizations, activists who had worked with these or other organizations, attorneys and paralegals who handled Central Americans' immigration cases, and Central Americans with recent or pending immigration cases. It was not difficult to gain access to the first three of these groups. Like many nonprofit agencies, Central American community groups in Los Angeles rely on the assistance of supporters who attend events, donate money, and volunteer their time. I had been involved in the Central American solidarity movement since the mid-1980s (see Coutin 1993), expressed sympathy with organizations' causes, and was willing to do volunteer work, so I was welcomed by organizations' staff. Moreover, one goal of these organizations is to disseminate information. This goal is usually accomplished through press conferences, radio announcements, and public meetings and trainings; however, speaking to researchers such as myself could also be considered a means of advancing this goal. Because community organizations compete for funding, prestige, and political space, the fact that I wanted to work with more than one community organization did present a problem to some staff members.[9] To address this concern, I agreed not to discuss the internal affairs of one group with members of another. Community activists and legal service providers were usually willing to grant interviews. In many cases, I met these individuals at community events, so by the time I asked for an interview, they already knew who I was and what I was doing. These individuals also knew each other, so using a snowball technique, I was able to meet and interview former activists. Many activists and legal service providers were used to being interviewed by journalists and researchers, and even used terms like "on the record" and "off the record" as they described their experiences. In short, given the nature of these individuals' work and the extent of scholarly and media interest in immigration, my own efforts to secure interviews were part of a broader and—to interviewees—familiar practice.

Arranging interviews with Salvadorans and Guatemalans who were seeking legal status was more difficult. I met most of these individuals through community organizations, or through attorneys whose cases I was following. When attorneys put me in touch with their clients, the clients—who trusted their attorneys—were usually quite willing to talk to me. When I approached legalization applicants[10] directly, however, it was a different matter. I limited my interview requests to individuals whom I met in the course of volunteering and who were not trying to obtain pro bono legal representation from community groups. I excluded the latter because I feared that they would accede to the interview request in the hope that I could influence the organization to accept their case. Although I did a variety of volunteer tasks, the work that brought me into direct contact with organizations' clients was preparing asylum applications and filling out work permit renewal requests. Many such clients were apprehensive about the asylum process or their pending immigration cases, so in such instances I did not ask for interviews. When I met clients who seemed talkative, however, I explained that I was a professor engaged in research regarding the legal situation of Central American immigrants, and I asked if they would be willing to participate in an interview on a subsequent occasion. I stressed that the interview was entirely separate from the community organization, that the information would be anonymous, and that we could schedule the interview at a time and place that would be convenient to them. About twenty percent of these individuals declined, saying that they were too busy or that they didn't have anything to say. I suspect that in addition to their stated reasons, many were apprehensive about potential legal or political consequences. Most potential interviewees had lived through civil wars that were characterized by severe human rights violations. If they were applying for political asylum, then discussing their legal case could mean discussing politically sensitive information, or revisiting painful experiences. In addition, the Spanish language press continually reported on immigration raids at workplaces, cases of individuals being apprehended by immigration agents while asleep in bed, and children coming home from school to discover that their parents had been deported. Such enforcement practices, along with the possibility that policies would become even more restrictive, generated a climate of fear.[11] Given this climate, many individuals probably concluded that it was better not to discuss their legal situation with strangers.

Those individuals with pending immigration cases who did agree to participate seemed to see the interview as an opportunity to counter restrictionist immigration policies. Some such interviewees had attended universities in Central America and saw the interview as similar to research that they had conducted. Interviewees who had been active in political movements in Central America or who supported community organizations in Los Angeles characterized the interview as a form of "collaboration."

Almost all interviewees with pending cases spent part of the interview refuting allegations that immigrants commit crimes, take welfare, and steal jobs from U.S. citizens. For example, after he had answered all of my questions, I asked Salvador Carranza, a thirty-nine-year-old Salvadoran man, whether he had anything to add. He responded,

> One thing is that politics here are very inflammatory for people who come from other countries. They are proposing many laws against the Hispanic right now, though they aren't directly against the Hispanic, because they affect everyone, whether they're from Europe or from Asia or from elsewhere. But because our countries are so close together, more than anyone else, these laws affect the Hispanic. And there's a false image of immigrants now. All kinds of people have come here, and many are very prepared [well-educated]. And the costs of preparing them were born by their country, not by their country. People come here with the goal of working. We aren't accustomed to taking public aid. Everyone who I know from my country likes to work. Here, there are discriminatory people, racist people. The people who are proposing these laws are motivated, probably, by fear, the fear that a minority will become a majority. The fear of losing power. And this isn't right because everyone who comes here works, everyone pays taxes.

Although I did not question them about the validity of anti-immigrant stereotypes, interviewees like Salvador perceived the interview itself as posing an implicit question about the legitimacy of their presence. Despite the fact that I did not consider myself the sort of person to whom it would be necessary to rebut stereotypes, interviewees' statements positioned me as a representative of Anglo U.S. society through whom they could counter anti-immigrant policies.

Establishing "Rapport"
The politics of language, race, class, and to a lesser extent, gender also influenced my interviews and fieldwork. I considered Spanish-language skills essential to conducting research among Central American immigrants, many of whom were primarily Spanish-speaking. Yet, the political and legal context gave additional meanings to the act of conducting an interview in Spanish.[12] On the political front, advocates of immigration reform complained that immigrants were unwilling to assimilate, learn English, and adopt U.S. customs. In addition, the Unz initiative, which dismantled bilingual education in the state of California, depicted multilingualism as an obstacle to educational achievement. Given this political climate, it could be considered an act of solidarity for an Anglo English-speaker to conduct an interview in Spanish. On the legal front, many interviewees had been advised that learning English would help their immigration cases. Immigration judges consider English fluency to be a sign that immigrants have set down roots in the United States, increasing the hardship that a deportation would impose (Coutin 2000). Immigration attorneys therefore advise clients who are able to do so to testify in English. Given this political and legal context, community organizations' clients were often surprised that an Anglo U.S. citizen like myself had bothered to learn Spanish. Speaking Spanish, which was to me both a practical necessity and a way of demonstrating my respect for interviewees, had a political significance that I did not intend.

In addition to language, race was a salient category during interviews. As mentioned previously, interviewees with pending legalization cases often spent part of the interview denouncing the racism of anti-immigrant politics. I sympathized with such remarks, and agreed with interviewees' often-expressed view that it was unfair to blame all immigrants for the actions of a few.[13] Some interviewees, however, spoke about race in ways that made me uncomfortable. For example, Ana Sanchez, a Salvadoran woman, stated that Latinos didn't spend enough time with their children and were therefore bad parents. Such pejorative generalizations contradicted my own sense that race is not a valid basis for making conclusions about people's behavior. Similarly, when another interviewee and I discussed the difficulty of parking near his apartment building, the interviewee commented that the parking laws are disobeyed by local residents. He said, "Someone of your race, when they see a parking sign will know that that means that you can't park there and that it has to be obeyed. But someone of my race will most likely ignore the sign and park there anyway." This remark attributed parking behavior to race, a connection that I found disconcerting. Moreover, the interviewee's use of the terms "your race" and "my race" emphasized racial distinctions over political commonalities. The unavoidable hegemony of racializing discourses was further conveyed through an anecdote that several interviewees related. I quote here the version told by Rodolfo Nuñez, a former student activist from El Salvador: "Do you remember the case of the mayor of Pomona last year? ... There was a, what are those called, an immigration raid in the highway, in the freeway. And it turned out that he still had his little old truck... and his Hispanic appearance. They stopped him, they detained him ... Afterwards they apologized when they realized that he was the mayor, the mayor of Pomona." The salience of race disrupted the facade of researcher and interviewee engaged in a common endeavor and exposed ways that both I and those I interviewed were implicated in the hierarchies that characterize immigration and identity politics.[14]

Interviewees also pointed out the intersection between race and class. I tried to deemphasize class distinctions between myself and those I was interviewing.[15] For instance, when women told me that they worked as maids or as housekeepers, I commented that I used to do such work myself. For instance, when Angela Reyes, a Salvadoran interviewee with a pending asylum application, told me that she cleaned hotel rooms for a living, I described my own experiences working as a maid at a Holiday Inn. Angela then said with surprise, "I thought only Latinas got those jobs." Similarly, when I met Juan and Maria Bonilla at their home for an interview, I commented that the lay-out of their apartment was identical to one that I'd lived in as a college student. Juan and Maria were surprised to learn that, like them, I had shared a one-bedroom apartment with three other people. Some interviewees interrogated my efforts to claim commonality. Rodolfo Nuñez, who I quoted above, argued that immigrants are not taking jobs away from citizens. "Would you be willing to work for $35 a day?" he asked me. When I told him that I used to work as a maid and as a janitor, he asked me when I had done such work. I admitted that I had done so temporarily, while working my way through school. Similarly, Deysi Mendoza, a Salvadoran asylum applicant, asked me, "Would you want to work for $50 a day, cleaning houses?" I told her that I had done this sort of work earlier in my life, but that I now had other opportunities. She then pointed out that I was able to pay someone else to care for my children but that she cared for other people's kids. "See?" she said, "White people don't want to do the jobs that we do." Racialized class differences entered interviews in more subtle ways, as well. Another interviewee concluded our interaction by asking me to recommend her to any friends who needed someone to clean their homes. Clearly, the idea that hierarchies can be temporarily suspended while researcher and researchee "establish rapport" highlights methodological dimensions of interaction to the exclusion of other dynamics. In contrast, Rodolfo Nuñez, Deysi Mendoza, and other interviewees did not see the "interview" as isolated from hierarchies that otherwise pervade our lives.[16]

Race and class also figured in an experience that I had several times during field-work: that of being mistaken for someone else. When I attended immigration court to observe a hearing, attorneys (who are often eager for information about policy, strategies, cases, and experiences) would lean over and ask, "Are you an attorney?" Apparently, the fact that I was White and dressed in professional attire (skirt and blouse or sweater) led them to reject the possibility that I was an immigrant with a pending case. Similarly, when I went to the Los Angeles Convention Center to attend a naturalization ceremony and asked security guards for directions, guards would ask me, "INS?" Seemingly, my appearance and accent distinguished me from naturalizing citizens and their relatives. Central Americans whom I met during field-work sometimes worried that I had connections with the Immigration and Naturalization Service. When legal staff at community organizations asked clients whether or not I could observe their conversations, clients sometimes asked, "But is she from the INS?" One interviewee told me that when she first entered the U.S., she thought that all Americans were from Immigration, especially people who were very, very white. "In those days," she said, "I would have hidden from you if I saw you. I would have been in a panic!" Even community activists seemed to feel that I looked the part of an INS official.[17] When one community organization considered making a training video to acquaint asylum applicants with the asylum process, staff asked me if I would play the role of an INS official questioning an applicant about his or her case. Again, my appearance, language, and knowledge positioned me within immigration hierarchies.

While I felt that my race and class posed problems that I had to overcome during fieldwork, I rarely perceived gender as an obstacle to fieldwork. I spent a lot of time talking to both men and women about their experiences as parents, in part because I was pregnant and then the parent of a newborn when I began fieldwork. It may be that gender shaped topics that interviewees were willing to discuss. Women, for example, talked to me about their fears of being raped while en route to the United States, their efforts to evade alien smugglers' sexual advances, their relationships with their spouses or partners, their concerns as mothers, and, more rarely, the ways in which they felt freer from gender constraints in the U.S. than they had been in El Salvador or Guatemala. Women might have been more reticent regarding these topics if they were talking to a man. It is difficult to say whether men would have discussed other topics if I had been male. One male interviewee did refer to himself as "machistic" in that he wanted to qualify for legal status on his own instead of through his wife, already a U.S. citizen. Gender was more of an issue when I tried to participate in networks of Salvadoran political activists. Many male activists held full-time jobs and spent their evenings and weekends working for a cause. I felt the need to demonstrate similar dedication, but I didn't want to be away from my children that long. Such gender divisions became visible at one rally that I attended. When a male activist called a congressional representative on a cell phone, the other male activists who were present grouped around the caller to learn the news and formulate a response. I found myself standing next to a Salvadoran woman activist who, like me, had not been included in this grouping. "Look," she commented, "All men!" She pointed out that these particular men had been raised in a political atmosphere, had dedicated themselves to these issues, and now depended on their wives to take care of their children. I noted that in my own case, childcare responsibilities made presence at evening and weekend political meetings difficult to achieve.

Presence

I was present at events for multiple reasons. For research purposes, I needed to observe and take notes on court hearings, attorney–client interviews, case preparation, community events, political meetings, strategy sessions, parties, picnics, vigils, and demonstrations. My presence at these events was never "just" or even mostly research, however. Because I volunteered at community organizations, my observations of staff also constituted a form of "training" that eventually enabled me to assess the cases of community organizations' clients myself. The fact that I was observing legal work and court proceedings also made me a "witness" with knowledge that could support or challenge particular legal claims. I formed ties of friendship and colleagueship with staff at community groups, so I attended events for social as well as research reasons. Politically, I sympathized with Central Americans' efforts to obtain permanent legal status in the United States, so I attended rallies not only to observe others, but also to express my own views. As I wrote in my fieldnotes regarding a week-long fast and vigil that occurred in Los Angeles in November 1997, just before Congress approved the Nicaraguan Adjustment and Central American Relief Act:

> This process has been interesting to me as a culmination of my fieldwork. I find that I know lots of the people who are participating in this activity (though the absences are also notable), and that I have a different role—writing press releases, attending coalition meetings, but never being challenged or called on in quite the same way that others are. My participation is sort of optional in [a] way that other people's isn't. But at the same time it isn't optional—this is a commitment I've made to people and that I have to act on. And I feel pulled in different directions—Am I compromising my status as a researcher by writing the press releases? Am I not doing enough to fulfill my commitment to others? And am I getting the sort of information that I should be getting?

Some might complain that such competing commitments made me biased and my results subjective. I admit my biases and do not claim objectivity. I nonetheless would also argue that, because social interaction is multifaceted, fieldwork unavoidably creates competing obligations of one sort or another (see Coutin and Hirsch 1998).

The political implications of my fieldwork presence were complicated. Attending a political demonstration signifies supporting a cause. In fact, activists measured the success of an event in part by the number of people who came. I also attended planning meetings and strategy sessions. Even if I had been there solely for research purposes, my presence would have acquired other meanings. Like other meeting participants, I signed attendance lists. At the end of each meeting, tasks were divided up, often by organization, but sometimes also by individual. I assisted with press releases. The volunteer work that I did at community organizations was useful for my research but was always also a way of supporting community organizations' work. Community organizations even asked for explicit statements of solidarity. One organization asked to include my name in an ad that criticized changes in U.S. asylum policy. I agreed to sign the ad. There were also complicated political rivalries between the three Central American community organizations with which I was working. I wondered why all three agreed to let me observe their activities, given that their relationships with each other were characterized by mistrust as well as collaboration. It's possible that they saw me as sympathetic, agreed that my research would be valuable, and basically trusted me. It's also possible that they agreed to participate because if they had not done so, only their "competition" would have been represented in the study. In other words, suspicion and rivalry as well as trust can facilitate research. Finally, my fieldwork elicited politically sensitive information from asylum applicants and Salvadoran activists. In the political realities of El Salvador and Guatemala, to admit that one has been a target of persecution is to depict oneself as politically suspect (Coutin 2001). As a result, talking about human rights abuses can jeopardize speakers' personal security.

My presence at community organizations, in court, and as an interviewer was also legally complex. When I volunteered at community organizations, I was engaging in legal activity. I translated documents, filled out forms, interviewed organizations' clients, and helped to draft asylum declarations. As I did this work, I was aware that knowledge can be legally dangerous. If an individual tells a legal worker something that conflicts with the individual's official account, then the legal worker cannot certify that this account is true to the best of his or her knowledge. Organizations also realized that they ran legal risks in permitting me to observe their activities. If I were to observe and document legally questionable activities, such as preparing frivolous asylum applications, then my testimony or my field data could incriminate them. (In fact, I did not notice any legally questionable actions on community organizations' parts.) When I observed court hearings, I also was a witness of sorts. Once, when I was waiting for an asylum hearing to begin, I observed the judge ask an attorney if the attorney's client was in court. The attorney responded that the client—an indigenous Guatemalan—had not yet arrived. "Maybe he went back to the mountains," the judge remarked. The attorney interpreted this remark as a sign of prejudice against his client. Because this remark was not on the record, the attorney asked me to sign an affidavit describing what I observed. My affidavit became part of an appeal of the judge's denial of this asylum petition. Interviews also elicited legally sensitive information. One interviewee lowered his voice when he told me that his wife was undocumented. Another interviewee stopped talking whenever two technicians who were installing cable in her home came within earshot of the kitchen table where we were speaking.

As the above examples show, "research" and "practice" overlapped considerably in my fieldwork. Eventually, I had observed so many interviews, case review meetings, and court hearings that, when they were short-staffed, attorneys asked me to interview their clients myself. I complied with such requests. Additionally, at case review meetings, legal staff began to ask my opinion about attorney–client interviews that

I had observed. At one meeting, an attorney wanted to know whether or not I had found a client credible. I answered that the client had seemed credible to me, but that I really could not judge. At another meeting, I used notes that I had taken while observing their interview to provide additional information about a client. Though they encouraged me to participate in this way, legal staff also remembered that I was a researcher. Once, while discussing a complicated legal dilemma with staff, an attorney commented, "I'm choosing my words carefully because I'm aware of the anthropologist." When this occurred, I immediately stopped taking notes. Despite the considerable overlap between "research" and "practice," there were occasions when I deliberately avoided becoming involved in organizations' work. For instance, two organizations considered a controversial proposal to merge their legal departments. An administrator at one of these organizations asked me to use my expertise regarding the "organizational culture" of each group to facilitate the merger. I declined, reasoning that this would violate the confidentiality agreements that had enabled me to work with multiple and sometimes competing groups.

As I helped Central Americans document their legal claims while I produced my own documentation of their efforts to secure legal residency, I noted the similarity between legal and ethnographic forms. Paralegals had to decipher clients' legal histories in order to devise effective legal strategies. I had to decipher these legal histories in order to understand interviewees' legal categories, strategies, and experiences. Legal staff created case files for every client that they represented. I maintained a record of every interview that I conducted. Community organizations required clients to sign contracts specifying their rights, obligations, and expected services. I presented interviewees with consent forms that detailed their rights as research subjects, the nature of the interview, and measures to minimize any potential risks. Through affidavits, check stubs, medical reports, letters, news articles, and other records, legalization applicants had to document their narratives of persecution and their lives in the United States. I used fieldnotes, interview transcripts, news clippings, and other documents to develop an account of Central Americans' legalization strategies. Clearly, the need to establish records, document claims, and assess truth are part of both law and ethnography. In this sense, both ethnography and law employ similar methods.[18]

Conclusion

Ethnography is never "just" ethnography. As research techniques, participant observation and interviews cannot be isolated from the context in which they occur. Methodological dilemmas are not limited to devising means of "gaining access" to people, information, and events; or to overcoming "obstacles" that interfere with the pursuit of knowledge; or to behaving in an ethical fashion while "in the field." In addition, ethnographers confront the fact that the production of knowledge is a political process that is beyond ethnographer's control. Like other researchers, my interest in Central Americans, immigration, and identity politics was informed by broader political debates regarding these topics. Recent theory regarding social subjectivity led me to focus on contexts in which identities were assigned, claimed, and contested. These contexts included attorney–client interviews, case review meetings, case preparation, court hearings, and political meetings and demonstrations. I was particularly interested in identifying the ways that the individuals who were being defined by immigration categories shaped the definitions that were being produced. This concern led me beyond the realm of official legal proceedings to other aspects of immigrants' experiences. In interviews, Salvadorans and Guatemalans described how they had transgressed and been confined by legal categories. As a shaping force, law was both irrelevant and all-too-real within their lives.

Although I had clear analytical goals in conducting interviews, observing events, and participating in activities, the political and legal context in which I did fieldwork transformed these research techniques into political and legal acts. I found that interviewees with pending legalization claims perceived interviews as a means of voicing their opposition to more restrictive policies. Interviews were directed not only by my analytical inquiry but also by public debate. Some interviewees saw participating in interviews as a means of collaborating with community efforts to assert rights as immigrants. The identity and immigration politics that pervaded public debate also infused interviews and fieldwork. Although I sympathized with immigrants politically, I was also distanced from them by legal, racial, class, and, less often, gender hierarchies during fieldwork. The fact that I was doing research did not erase the hierarchies that otherwise shape social relations. My presence also had political and legal ramifications. I learned politically and legally sensitive information, I took legal actions, I witnessed events, I took political stances, and I practiced what I was studying. My research therefore had multiple meanings.

If ethnographic methods are redefined by the context in which they occur, then ethnography is continually being reinvented and reconceptualized. Practices endure, but the situations in which these practices are employed undergo change. In the years since Malinowski set out to study Trobriand Islanders, ethnographers have turned their attention to nuclear weapons manufacturers (Gusterson 1996), new reproductive technologies (Martin 1987), and the internet to name but a few topics. The versatility and analytical depth of ethnographic methodologies has drawn attention within disciplines outside of anthropology. In devising ethnographic projects, it is important to bear in mind that ethnography is not a "tool kit" from which a researcher can simply extract and apply research techniques. Rather, ethnography entails devising complex relationships with people who are themselves positioned within social institutions, groupings, and processes. Using these relationships to produce knowledge has political and other implications that are beyond ethnographers' control. It is important for researchers to consider these implications, not only when writing but also when conceptualizing a project. Only thus can ethnography be revealed as the rich and complicated process that it is.

Notes

I am grateful to June Starr for organizing the sessions for which I wrote the original versions of this chapter, and to both June Starr and Mark Goodale for their insightful comments. This chapter benefited from my conversations with Nahum Chandler, Sue Hirsch, and Bill Maurer. Sue Hirsch also provided comments on an earlier draft. My fieldwork in Los Angeles was funded by the National Science Foundation (grant #SBR-9423023). I am indebted to ASOSAL (the Association of Salvadorans of Los Angeles), CARECEN (the Central American Resource

Center), CHIRLA (the Coalition for Human Immigrant Rights of Los Angeles), the Coalición Centroamericana, El Rescate, and the many individuals who agreed to participate in interviews.

1. For critiques and rethinkings of ethnographic representations, see Asad 1973; Clifford and Marcus 1986; Gupta and Ferguson 1997; Marcus and Fischer 1986; Narayan 1989; Rosaldo 1989 and Starn 1994.

2. The Antiterrorism and Effective Death Penalty Act, P.L. 104-132 (110 Stat. 1214); Illegal Immigration Reform and Immigrant Responsibility Act. P.L. 104-208 (H.R. 3610), September 30, 1996; Personal Responsibility and Work Opportunity Reconciliation Act of 1996. P.L. 104-193 (110 Stat. 2105).

3. At the same time, critiques of the ways that "identities" were being defined in public debates emerged from other quarters, among activists who considered such notions to be overly focused on "problems of individual awareness, communication, and sensitivity" (Gregory 1994: 21) and therefore themselves imbued with hegemonic and racist implications (Gilroy 1987).

4. On increasing racial and economic segregation within Los Angeles, see Davis 1992.

5. Probably the most important factor, however, was my prior work regarding the U.S. sanctuary movement. See Coutin 1993.

6. One of colonialism's legacies, for example, has been the persistence of discourses of "othering" that distinguish between primitive and modern, exotic and normal, the cultural and the cultureless, the East and the West, the timeless and the historical (Price 1983; Rosaldo 1989; Said 1979; Shapiro 1988; Taussig 1987). Similarly, subordinating categories of personhood mark groups for particular positions within the labor force, as demonstrated by the feminization of certain forms of production (Arizpe and Aranda 1981; Ong 1987) and the reliance upon undocumented labor within certain industries (Bach 1988; Jenkins 1978; Portes 1978; Sassen 1989).

7. I do not mean to imply that there are no differences between the social position occupied by the documented and the undocumented. Obviously, undocumented individuals are concentrated in particular industries and neighborhoods. Nonetheless, these jobs and individuals are shared with citizens and legal permanent residents.

8. I think, for example, of Laura Bohannan's (1954: 123) discussion of her decision to dance with (presumably Tiv) women during a wedding: "If I were to dance at all, I had to concentrate on the music and my muscles, but while I danced, my anthropological conscience nagged that I was missing something."

9. For another researcher's account of her efforts to negotiate these tensions, see Baker-Cristales 1999.

10. I use the term "legalization applicant" because the Central Americans whom I interviewed applied for legal status through a variety of mechanisms, including political asylum, a family visa petition, an employer petition, the diversity visa program, and suspension of deportation. Technically, however, they were not applying for *legalization* but rather for the specific benefit (such as asylum or suspension) that the legalization mechanism could confer. In contrast, during the 1986 "amnesty" program, individuals applied directly for legalization.

11. In fact, both interviewees and individuals whom I met while volunteering sometimes said that they hadn't applied for legal status earlier because they were afraid of revealing their identities to the INS.

12. My initial conversations with many community organizations' clients focused on language, with clients asking how and where I had learned Spanish. When I explained that I had studied Spanish and traveled in Latin America, some expressed mild embarrassment that, after living in the United States for five years or more, they had not yet mastered English. In contrast, I always felt embarrassed that my Spanish language skills were not better.

13. Actions that interviewees depicted as negative included not working, receiving welfare, and committing crimes. I did not necessarily see all of these things as negative.

14. These hierarchies were most apparent in interviews with legalization applicants. Attorneys, legal staff, and community activists were situated differently vis á vis these hierarchies than were their clients.

15. Class differences were most pronounced when I was interviewing individuals with pending legalization cases. Many such interviewees worked in blue collar jobs or in the informal economy. In contrast, activists and legal service providers were professionals.

16. In other words, they didn't simply argue that immigrants weren't taking jobs away from citizens, they also argued that they hadn't taken a potential job away from me.

17. In making this request, staff were not only thinking of my appearance. They also knew that I had attended asylum interviews as an interpreter, and was therefore familiar with interview procedures.

18. I do not mean to imply that ethnography and law are identical. My own assessment of cases and credibility often differed from that of immigration judges, for example.

References

Anderson, Benedict
    1991   Imagined Communities: Reflections on the Origin and Spread of Nationalism. (Rev. ed.) London: Verso.
Anzaldua, Gloria
    1987   Borderlands/La Frontera. San Francisco: Spinsters/Aunt Lute.
Appadurai, Arjun
    1991   "Global Ethnoscapes: Notes and Queries for a Transnational Anthropology." In Richard G. Fox, ed., Recapturing Anthropology: Working in the Present, pp. 191–210. Santa Fe: School of American Research Press.
Arizpe, Lourdes and Josefina Aranda
    1981   "The 'Comparative Advantages' of Women's Disadvantages: Women Workers in the Strawberry Export Agribusiness in Mexico." Signs 7: 453–473.
Asad, Talal, ed.
    1973   Anthropology and the Colonial Encounter. London: Ithaca Press.
Bach, Robert L.
    1978   "Mexican Immigration and the American State." International Migration Review 12(4): 536–558.
Baker-Cristales, Beth
    1999   "Politics and Positionality in Fieldwork with Salvadorans in Los Angeles." PoLAR: Political and Legal Anthropology Review 22(2): 120–128.
Bohannan, Laura (a.k.a. Elenore Smith Bowen)
    1954   Return to Laughter. New York: Doubleday.
Butler, Judith
    1993   Bodies that Matter: On the Discursive Limits of Sex. New York: Routledge.
Clifford, James
    1988   "Identity in Mashpee." In The Predicament of Culture: Twentieth-Century Ethnography, Literature, and Art, pp. 277–348. Cambridge, MA: Harvard University Press.
Clifford, James, and George E. Marcus (eds.)
    1986   Writing Culture: The Poetics and Politics of Ethnography. Berkeley: University of California Press.
Collier, Jane, Bill Maurer and Liliana Suraez-Navaz.
    1995   "Sanctioned Identities: Legal Constructions of Modern Personhood." Indentities 2(1–2): 1–27.
Coombe, Rosemary J.
    1998   The Cultural Life of Intellectual Properties: Authorship, Appropriation, and the Law. Durham, NC: Duke University Press.

Coutin, Susan Bibler
  1993   The Culture of Protest: Religious Activism and the U.S. Sanctuary Movement. Boulder, CO: Westview Press.
Coutin, Susan Bibler
  1994   "The Negotiation of Legal Identity among Salvadoran Immigrants in Los Angeles, California." Grant Proposal submitted to the National Science Foundation.
Coutin, Susan Bibler
  1998   "From Refugees to Immigrants: The Legalization Strategies of Salvadoran Immigrants and Activists." *International Migration Review* 32(4): 901–925.
Coutin, Susan Bibler
  2000   Legalizing Moves: Salvadoran Immigrants' Struggle for U.S. Residency. Ann Arbor: University of Michigan Press.
Coutin, Susan Bibler
  2001   "The Oppressed, the Suspect, and the Citizen: Subjectivity in Competing Accounts of Political Violence." *Law and Social Inquiry* 26(1): 63–94.
Coutin, Susan Bibler and Phyllis Pease Chock
  1995   "Your Friend, the Illegal: Definition and Paradox in Newspaper Accounts of U.S. Immigration Reform." *Identities* 2(1–2): 123–148.
Coutin, Susan Bibler and Susan F. Hirsch
  1998   "Naming Resistance: Ethnographers, Dissidents, and States." *Anthropological Quarterly* 71(1): 1–17.
Davis, Michael
  1992   "Forgress Los Angeles: The Militarization of Urban Space." In Michael Sorkin (ed.), Variations on a Theme Park: The New American City and the End of Public Space, pp. 154–180. New York: Hill and Wang.
Derrida, Jacques
  1976   Of Grammatology, G. Spivak, trans. Baltimore, MD: Johns Hopkins University Press.
Engel, David
  1993   "Origin Myths: Narratives of Authority, Resistance, Disability, and Law." *Law and Society Review* 27(4): 785–826.
Evans, Sara
  1979   Personal Politics: The Roots of Women's Liberation in the Civil Rights Movement and the New Left. New York: Vintage Books.
Feldman, Allen
  1991   Formations of Violence: The Narrative of the Body and Political Terror in Northern Ireland. Chicago: University of Chicago Press.
Foucault, Michel
  1979   Discipline and Punish; The Birth of the Prison. Alan Sheridan, trans. New York: Vintage Books.
  1980   The History of Sexuality, Vol. 1, An Introduction. Robert Hurley, trans. New York: Vintage Books.
Gilroy, Paul
  1987   "There Ain't No Black in the Union Jack": The Cultural Politics of Race and Nation. Chicago: University of Chicago Press.
Gordon, Colin, ed.
  1980   Power/Knowledge. New York: Pantheon.
Gregory, Steven
  1994   "We've Been Down This Road Already." In Steven Gregory and Roger Sanjek (eds.), Race, pp. 18–38. New Brunswick, NJ: Rutgers University Press.
Gupta, Akhil and James Ferguson
  1997   "Beyond 'Culture': Space, Identity, and the Politics of Difference." In Akhil Gupta and James Ferguson (eds.), Culture, Power, Place: Explorations in Critical Anthropology, pp. 33–51. Durham, NC: Duke University Press.

Gusterson, Hugh
    1996   Nuclear Rites: A Weapons Laboratory at the End of the Cold War. Berkeley: University of California Press.
Hooks, Bell
    1984   Feminist Theory from Margin to Center. Boston: South End Press.
Heyman, Josiah McC.
    1998   Finding a Moral Heart for U.S. Immigration Policy: An Anthropological Perspective. Arlington, VA: American Anthropological Association.
Jenkins, J. Craig
    1978   "The Demand for Immigrant Workers: Labor Scarcity or Social Control?" *International Migration Review* 12(4): 514–535.
Kristeva, Julia
    1981   "Woman Can Never Be Defined." In Elaine Marks and Isabelle de Courtivron (eds.), New French Feminisms, pp. 137–141. New York: Schocken.
Macpherson, C. B.
    1962   The Political Theory of Possessive Individualism: Hobbes to Locke. Oxford: Clarendon Press.
Malinowski, Bronislaw
    1961 [1922].   Argonauts of the Western Pacific. New York: Dutton.
Malkki, Liisa H.
    1992   "National Geographic: The Rooting of Peoples and the Territorialization of National Identity Among Scholars and Refugees." *Cultural Anthropology* 7(1): 24–44.
Marcus, George and Michael M. J. Fischer
    1986   Anthropology as Cultural Critique: An Experimental Moment in the Human Sciences. Chicago: University of Chicago Press.
Martin, Emily.
    1987   The Woman in the Body: A Cultural Analysis of Reproduction. Boston: Beacon Press.
Martin, Philip
    1995   "Proposition 187 in California." *International Migration Review* 29(1): 255–263.
Matsuda, Mari J., Charles R. Lawrence III, Richard Delgado, and Kimberle Williams Crenshaw
    1993   Words that Wound: Critical Race Theory, Assaultive Speech and the First Amendment. Boulder, CO: Westview Press.
Merry, Sally
    1990   Getting Justice and Getting Even: Legal Consciousness among Working-class Americans. Chicago: University of Chicago Press.
Narayan, Kirin
    1989   Storytellers, Saints, and Scoundrels: Folk Narrative in Hindu Religious Teaching. Philadelphia: University of Pennsylvania Press.
Nordstrom, Carolyn, and Antonious C. G. M. Robben
    1991   Fieldwork Under Fire: Contemporary Studies of Violence and Survival. Berkeley: University of California Press.
Ong, Aihwa
    1987   Spirits of Resistance and Capitalist Discipline: Factory Women in Malaysia. Albany: State University of New York Press.
Perea, Juan F., ed.
    1997   Immigrants Out! The New Nativism and the Anti-Immigrant Impulse in the United States. New York: New York University Press.
Portes, Alejandro
    1978   "Toward a Structural Analysis of Illegal (Undocumented) Immigration." *International Migration Review* 12(4): 469–484.

Price, Richard
    1983   First-Time: The Historical Vision of an Afro-American People. Baltimore, MD: Johns Hopkins University Press.
Rosaldo, Michelle and Louise Laphere (eds.)
    1974   Woman, Culture, and Society. Stanford, CA: Stanford University Press.
Rosaldo, Renato
    1989   Culture and Truth: The Remaking of Social Analysis. Boston: Beacon Press.
Said, Edward
    1979   Orientalism. New York: Vintage Books.
Sassen, Saskia
    1989   "America's Immigration 'Problem': The Real Causes." *World Policy Journal* 6(4): 811–831.
Shapiro, Michael J.
    1988   The Politics of Representation: Writing Practices in Biography, Photography and Policy Analysis. Madison: University of Wisconsin Press.
Starn, Orin
    1994   "Rethinking the Politics of Anthropology: The Case of the Andes." *Current Anthropology* 34: 13–38.
Taussig, Michael
    1987   Shamanism, Colonialism, and the Wild Man: A Study in Terror and Healing. Chicago: University of Chicago Press.
White, James Boyd
    1985   Heracles' Bow: Essays on the Rhetoric and Poetics of the Law. Madison: University of Wisconsin Press.
Yanagisako, Sylvia and Carol Delaney (eds.)
    1995   Naturalizing Power: Essays in Feminist Cultural Analysis. New York: Routledge.
Yngvesson, Barbara
    1993   Virtuous Citizens, Disruptive Subjects: Order and Complaint in a New England Court. New York: Routledge.
    1997   "Negotiating Motherhood: Identity and Difference in 'Open' Adoptions." *Law and Society Review* 31(1): 31–80.

CHAPTER 7

ETHNOGRAPHY IN THE ARCHIVES

Sally Engle Merry

An anthropologist using archival legal texts, such as court records, faces a challenging situation: In order to make sense of these records, she has to ground them in an ethnography of the surrounding community. This includes an analysis of its major actors and its changing political, economic, social, and cultural terrain over a long period of time. In order to do this, the anthropologist must do ethnography in the archives. For an anthropologist like me who had previously done only contemporary ethnography, this was an intriguing and painstaking project, but one that was very rewarding. I found myself able to talk about change over time far more easily, although the inability to discuss these changes with my subjects or to ask them any questions was always frustrating. I had to make do with the small clues they left behind. In this chapter, I will describe my efforts to do historical ethnography using nineteenth-century court records and surrounding archival information from a small town in Hawai'i. My argument is that understanding valuable primary records such as court dockets demands extensive attention to archival ethnography in the surrounding community.

My archival project developed out of an earlier ethnographic study of lower courts and mediation programs in Massachusetts that explored the way individuals used the law and the kinds of legal consciousness that inspired them to frame their family and neighborhood problems in legal terms (Merry 1990). I wanted to study local legal consciousness in a context in which the lower court law and everyday social practices in families and communities differed more dramatically than they did in Massachusetts. Colonialism is one, but only one, of the situations in which such disparities develop. Contemporary globalization and the expansion of the rule of law is another. My examination of court records from the period when the United States was colonizing Hawai'i showed law to be a critical mechanism of social transformation as well as a mark of civilization to the imperial powers (Merry 2000).

I have also done extensive contemporary ethnography on the same town in Hawai'i. I am currently writing a book on legal consciousness and the mobilization of legal sanctions in domestic violence cases in the 1990s in Hilo. During the late 1980s and 1990s, the number of cases in the courts concerning domestic violence expanded dramatically. This is a contemporary example of the sharp changes in the frequency of case types I observed in the nineteenth century.

This book on domestic violence cases in Hilo today also draws on historical information on domestic violence cases to emphasize the difference in the way courts now deal with domestic violence. Thus, my archival ethnography is sandwiched between two more typical ethnographic studies.

The archival part of the research began from a series of court record books called minute books, from a lower criminal court in a plantation and port town on the windward side of the island of Hawai'i. These minute books were preserved in a continuous sequence from 1852 to 1913. When I first looked at these books, I was fascinated by the access they provided to the everyday life of common people in Hilo in the nineteenth century. At the same time, I realized that it was not going to be easy to decide what they meant. In order to make sense of these detailed cases, it was necessary to engage in a much broader ethnographic study of the courts, community, and culture of this area. I needed to know who the judges and attorneys were, how cases came to court, what the dominant concerns about criminality and social justice were, and how the economic and political situation of the town had changed over this time period.

Although I had hoped to avoid it, I found that it was also necessary to examine changes in the legal system and court structure during this time period. Between 1825 and 1852 the independent Kingdom of Hawai'i appropriated large sections of Anglo-American law and legal practice, which substantially displaced Hawaiian law and practice. During this period, the Kingdom adopted a Constitution and Bill of Rights; an independent judiciary; a system of courts with lower courts, circuit courts, and a supreme court; written laws drawn largely from Anglo-American prototypes; and private landownership (see Merry 2000). When the islands were annexed by the United States in 1898, the legal system was already very similar to that of the U.S. American lawyers were instrumental in creating an American legal system in Hawai'i during this period. Thus, a substantial part of my research involved an examination of how this change came about and the complex set of steps whereby an autonomous kingdom adopted a foreign system of law and governance in order to preserve its sovereignty.

My historical study was resolutely ethnographic, exploring social factors such as the close network of social relationships joining judges, attorneys, and missionaries based on marriage, kinship, and shared school experience. I also examined the complex diversity of cultural assumptions shared by members of different ethnic groups that could be gleaned from private letters, journals, and unpublished documents as well as more public documents such as local newspaper stories and the texts of the laws. I relied to some extent on secondary historical sources that described the changing political economy of the islands during the nineteenth century. As a person who had previously done only contemporary ethnography, I was pleased to discover how the longer time span archival data provided allowed me to ask questions about change; this is normally difficult to approach in contemporary research. At the same time, I was frustrated by my inability to ask questions that were not included in the archives that had been preserved. In this chapter I will describe my approach to doing ethnography in the archives, indicating its possibilities as well as some of the limitations of this mode of anthropological inquiry.

Studying Court Records

I began with a large body of court records of the lower courts, called district courts, and the higher courts, called circuit courts. These courts were modeled after American prototypes rather than Hawaiian ones and are identified by English names. The court records included detailed descriptions of each case, including the names of the parties, the charge, the evidence presented, and the disposition. With the invaluable help of two research assistants, Marilyn Brown and Erin Campbell, I collected the texts of interpersonal cases from the district court of Hilo during this time span and tabulated the basic information on a sample of cases to enable statistical analysis of the cases. This project built on a previous statistical survey of all the Circuit Court cases in the Hawaiian Kingdom from 1849 to 1892 carried out by Harry Ball and Jane Silverman. I used the same categories for my data collection as this earlier study in order to produce comparable data from the District Court. But I focused only on the small port town of Hilo rather than the entire Kingdom of Hawai'i, in order to contextualize these cases within the social fabric of the town. The district court cases stretched from 1853 to 1910, after which the minute books were not preserved. I believe Harry Ball was responsible for rescuing the earlier books and having them archived in the Hawai'i State Archives in Honolulu. There is often considerable serendipity and agency in what gets preserved and what is lost.

I tabulated all the criminal cases for a full year once a decade from 1853 to 1903 in the District Court, a total of 2,325 cases.[1] In addition, I tabulated all the Circuit Court cases for a year every decade from 1905 through 1985 in the Circuit Court, a total of 817 cases.[2] I also analyzed Circuit Court cases in Hilo from 1852 to 1892 from the massive project supervised by Jane Silverman, with subsequent analysis by Harry Ball and Peter Nelligan.[3] I extracted the criminal Circuit Court cases from the Hilo region out of this data set, a total of 2,530 cases over the forty-year period. Taken together, I had statistical information on 5,672 district and circuit court criminal cases from Hilo and the surrounding region. Harry Ball and Jane Silverman generously shared the data from their previous research project with me.

Judges in the Hilo District Court, as in other district courts designated "police courts," were required to preserve "in written detail the minutes and proceedings of their trials, transactions, and judgements" according the Organic Acts of 1847 that established them (Statute Laws 1847: 12; Compiled Laws 1884: 237). The Hilo district court was designated a police court because of its added responsibilities in a harbor town where cases involved foreigners and mercantile issues as well as Native Hawaiians. For each case I recorded charge, plea, conviction, disposition, presence of an attorney, and gender and ethnicity of the defendant. I also collected the texts of all cases involving interpersonal relationships in the sampled years. In addition, I recorded all domestic violence cases that appeared in the district court for the entire time period so that I could track over time who returned to court with this problem. For about half of the period from 1853 to 1903, district court records were recorded in Hawaiian; the rest were in English. The Hawaiian records were ably translated by Esther Mookini, an experienced translator of nineteenth-century Hawaiian court records.

My project was to figure out who were in the courts as defendants, judges, and attorneys and to discover their social identities and connections. I tried to find out who married whom, who lived near whom, and who visited and went to parties together. One valuable source was a scrapbook of newspaper clippings about Hilo from the nineteenth century, collected by Luther Severance, a prominent member of the legal community, preserved in the Lyman House Memorial Museum in Hilo. Severance cut out articles from the Honolulu newspapers relevant to Hilo and carefully glued them into a scrapbook. Another was an autobiography written in Hawaiian and recently translated into English by a prominent Hilo lawyer, Joseph Nawahi (Sheldon 1909). In Hawai'i, I read numerous letters, journals, newspaper articles, and government documents collected in the Hawai'i State Archives, the Hawaiian Mission Children's Society Library, the Lyman House Memorial Museum in Hilo, and the libraries of the University of Hawai'i at Manoa and at Hilo. June Humme, a descendant of a prominent legal family in Hilo, generously shared her collection of articles, letters, and newspaper clippings about this family. I located two valuable unpublished reports (Kelly, Nakamura, and Barrere 1981; George 1948), several retrospective newspaper accounts (e.g., Lydgate 1922), and one master's thesis (Leithead 1974) written about the history of Hilo. Some of these were in the University of Hawai'i at Hilo archives, others in the University of Hawai'i at Manoa library or the Hawai'i State Library. I also looked at numerous documents from the sugar plantations of the area archived at University of Hawai'i at Hilo. These were supplemented by information on the sugar plantations at the Hawaiian Sugar Planters' Association archives in Aiea, near Honolulu.

In New England I found missionary letters and other documents at the Houghton Library at Harvard University; the American Antiquarian Society in Worcester, Massachusetts; the Peabody/Essex Museum in Salem, Massachusetts; and the Kendall Whaling Museum in Sharon, Massachusetts. My research included a substantial collection of magazine articles about Hawai'i in the popular and elite press of the nineteenth century, gathered from both the Wellesley College Library and the Hawai'i State Library by a research assistant, Nancy Hayes. I had access to the extensive resources of the Harvard University libraries as a Bunting Fellow. The Harvard Law Library and Widener Library held documents relating to the early nineteenth century, when the Kingdom of Hawai'i had important commercial and legal links with New England. Obviously, archival research requires a good deal of traveling.

In sum, I relied on a broad range of archival materials to learn about the small social world of this town in the nineteenth and early twentieth centuries. Since I do not read Hawaiian, I was unable to use Hawaiian-language sources, but I did consult recent scholarship based on this literature (e.g., Kame'eleihiwa 1992; Osorio 1996; Silva 1997). I supplemented the archival research with interviews of older residents of Hilo, including a man who played a prominent role in a bloody 1938 strike, Bert Nakane; and a man who served as a Democratic county attorney in the 1930s when the Republicans controlled the islands, Martin Pence. I also interviewed people involved with the plantation economy as early as the 1920s and older women who had grown up on plantations and could talk about the problem of wife beating in the camps and the indifference of the police. Obviously, this material does not stretch into the nineteenth century, but these interviews at least provided an ethnographic moment: an opportunity to pose questions. Together with the archival materials, they provided ethnographic context for the court records. I was able to establish who the judges and lawyers were and their relationships with the clientele of the courts. The close-knit social life of the Hilo elites and the deep chasm in social class, cultural understandings, and economic power between the judges and the defendants was particularly striking.

Obviously, not all groups in Hilo were equally represented in these archival accounts. While the elites wrote copiously about themselves and their lives—both whites and Native Hawaiians—the poor were notably silent. A few clues appeared: A protest against a boss on a plantation made it into the newspaper, or the report of high infant mortality rates among Filipino plantation workers in the early twentieth century engendered a study of their living conditions. A prominent spokesmen for the immigrant Japanese community was lynched in 1889, leading to a trial. But by and large, the underclasses are archivally silent. Their appearance in court records is a notable exception to their documentary erasure. It is for this reason that court records are such a valuable resource: It is only here that the stories and conditions of life of the poor, the marginal, and the non-literate appear.

Court Records in Nineteenth-Century Hilo

Much of the criminal work of the district and circuit courts of Hilo concerned infractions that were part of everyday social life: sexual activities, hitting, drinking, festivals such as cockfighting and gambling, and violations of work obligations. There was clearly a shift over time from a preoccupation with sex to drinking and drugs, gambling, and violations of the contract labor law. There was a shift from cases focusing on the family life and sexual behavior of Native Hawaiians to those concerned with the regulation of laborers on the sugar plantations (see Tables 7.1 through 7.6). Throughout the period, court cases benefiting the economic interests of judicial and plantation elites were embedded in a matrix of reformist cases. The naked economic functions of the legal system—privatizing land ownership and regulating labor—were clothed in the raiments of moral betterment.

A few plantations were established in the 1850s, but the massive expansion followed the Reciprocity Treaty of 1876, which gave Hawaiian sugar duty-free access to the American market. The declining Hawaiian population failed to provide enough workers to hoe and harvest the cane, leading to massive importation of labor under the contract labor system. Workers signed contracts to work for three years and were subject to penal sanctions if they refused to work. Importation of workers increased dramatically in the 1880s and 1890s, primarily from China but also from Portugal and, in the mid-1890s, Japan. At the turn of the twentieth century, some workers were brought from Puerto Rico, Korea, and, by the early 1900s, the Philippines. After the expansion of the plantations, the clientele in court consisted largely of recently arrived labor migrants. Although they came from different parts of the world, at each period their forms of recreation were typically criminalized, as was their resistance to working the long and arduous hours demanded of plantation labor. Thus, the people subjected to legal surveillance for these everyday offenses were mostly Native Hawaiians in the 1850s to 1870s, and largely immigrant plantation workers in the 1880s to 1900s (see Tables 7.1 through 7.6). Each wave of imported Asian and European laborers appeared in court in large numbers during the period of new immigration. However, as the immigrants from each country became established and formed communities and labor associations, they began to disappear from the court dockets.

Thus, defendants were disproportionately strangers, those new to the community and culturally outsiders to its emerging social order. In such a rapidly changing and culturally plural situation, the law served as the initial blunt instrument for cultural transformation. It was the method by which Native Hawaiians were molded into modern citizen subjects and stranger laborers were converted into a disciplined and docile labor force. Those running the courts and police, on the other hand, were established whites and Christianized Hawaiians. By the 1850s, these groups represented the old guard. As the century progressed, people of this background retained their control over the courts but gradually began to lose economic power as the sugar plantations increasingly fell under the control of outside, sometimes U.S.-based, mercantile interests that provided credit and managed the shipping and processing of the raw sugar. Meanwhile, the population of the town changed dramatically. The defendant population was increasingly made up of cultural "others." The courts were organized stratigraphically, with the oldest residents in charge and the more recent arrivals subject to their judicial decision making. This stratigraphic pattern has continued into the twentieth century as Japanese-Americans have become the core of the judiciary and court staff, along with whites and Hawaiians, while recent arrivals such as Tongans, Samoans, and Mexicans populate the defendant categories.

Nevertheless, class cross-cuts this stratigraphy in important ways. As long-term resident populations fall into the lower socioeconomic positions, they also become the object of legal attention. Thus, as one examines the shifting defendant population in this court over time, it is clear that the law serves as a front-line social control system, correcting the behavior of strangers to the social order, but over time, other mechanisms of ordering such as family, community, and workplace, come to take more critical roles in the control of everyday life. These more established groups dropped out of the defendant population. However, the focus of court surveillance continues to be the poor, of whatever ethnic group happens to occupy this social position.

Two of my central questions were (1) who were the judges and attorneys between 1850 and 1900 in this small town? and (2) who were the plaintiffs and defendants? Both the District Court and Circuit Court justices came from a small, predominantly white elite closely connected through social life and marriage. Most owned plantations and worked as attorneys in addition to their judicial duties, which were not full time until the end of the century. Many were politically active, serving in the House of Representatives: the lower legislative branch, modeled after that of the United States. They also had close ties to the missionary community, either through descent, marriage, or training. Virtually all were long-term residents of Hilo. According to an 1891 map of Hilo, all owned substantial house lots and houses in town. All were Christian and belonged to the Haili Church, formerly called the "foreign" church, housed in a wood-frame building reminiscent of Vermont churches. They represented "respectable" Hilo society, predominantly white but including a few mission-educated Native Hawaiian and Hawaiian-Chinese families. In some cases, the judge would face two attorneys, one of whom was his brother and the other his patron, each a member and descendant of the missionary community.

One of early judges was D. H. Hitchcock, the son of a missionary to Hawai'i, who was educated at Williams College and owned a plantation with his brother, who often represented clients in D. H.'s court.

One of the most long-lasting judges was George Washington Akao Hapai, whose name reveals his hybrid ancestry. Hapai was the District Court judge in Hilo for thirty years, from 1878 until 1908. His father was Chinese and had come to Hawai'i to perform the skilled work of making sugar for one of the Hawaiian monarchs in the 1840s. He married a high-ranking Hawaiian woman and owned a store in Hilo. I do not know if Hapai spoke Chinese, which is important since a substantial proportion of the defendants in the 1880s were from China. He was clearly most comfortable in Hawaiian and could speak some English. The court records he wrote are in Hawaiian with English legal terms. He was a Christian, a member of the Haili Church, to which the elite whites also belonged. His father was a Roman Catholic. Hapai was educated in a Protestant mission school. It seems likely that his cultural world was quite remote from that of the Chinese defendants.

Hapai was very closely connected to the white New England missionary elite of Hilo, particularly Luther Severance, a New Englander who occupied a prominent place in Hilo legal life for the last quarter of the nineteenth century. It appears that Luther Severance helped Hapai to get the position as judge since they appear to have worked together in the post office early in their careers. Unlike most of the white elites of Hilo, Hapai did not own a sugar plantation, but Luther Severance did. Severance came from Augusta, Maine. His family held politically powerful positions in Hawai'i and in the United States. Severance's father was a U.S. congressman for four years and was succeeded in Congress by James Blaine, who worked in Severance's father's newspaper office in Augusta, and was later Secretary of State of the United States. Severance was married to the daughter of a missionary to Hawai'i. His wife's sister was married to a circuit court judge in Hilo who also owned a sugar plantation. Severance served as Sheriff of Hawai'i from 1870 to 1884, a prominent position similar to that of a governor. In the 1880s Hapai frequently judged cases in which Severance served as prosecutor.

Another member of the judicial elite in Hilo was Joseph Nawahi, a politically prominent Native Hawaiian who appeared frequently in court as an attorney for the defense. He served several terms in the legislature. Nawahi was mission-educated and close to the one of the Hilo mission families as well as an active member of the Haili Church, but he spoke out on behalf of Native Hawaiians. As the local white elites turned from their support of the Hawaiian monarchy toward favoring some form of annexation to the United States in the late 1880s and early 1890s, Nawahi increasingly opposed them. He boycotted the church in 1890 when the trustees refused to let a supporter of the monarchy speak there.[4] According to his biography, Nawahi was born in 1842, attended the missionaries' school in Hilo and lived with a mission family for some time, then went to more advanced missionary schools in Lahaina and Honolulu to learn English (Sheldon 1909). Nawahi was a member of the educated, Christian Native Hawaiian elite and close to the missionary community until the politics of sovereignty drove them apart. His wife was a member of the elite Hawaiian-Chinese community in Hilo and also a member of the Haili Church (Kai 1976: 19). Nawahi frequently appeared in court defending Native Hawaiians and plantation workers charged with refusing to work or attacking their bosses. He increasingly opposed some of the missionary descendant lawyers who made up the old guard of Hilo society as the American-led annexation movement gained strength.

What are the implications of the structure of the local judiciary in Hilo? The striking feature of the pattern of court cases and defendant populations is the focus on social reform. The people running the courts tried to reform social behaviors they considered repugnant or harmful. They came from a Christian missionary tradition and brought to the judicial function a sense of responsibility to maintain marriages and prevent divorce, to control the consumption of alcohol and opium, to foster good habits of work and punctuality, to prevent violence and protect the public order, and to prohibit gambling. Their preeminent concern was the character of the Native Hawaiians, always the center of the missionary project. The moral character of the stranger populations brought to harvest the cane was of very little interest except when they were imagined to exert a destructive moral influence on the Hawaiians. Considerable legal attention was devoted to prosecuting people who sold alcohol to Hawaiians, for example. Few of the newcomers were prosecuted for adultery or lewdness; almost all the defendants were Native Hawaiians. Reforming Hawaiian marriages to convert them into lifelong and sexually exclusive relationships was long a goal of the missionary and legal communities that were so closely connected to one another. But the newcomers were potential workers, not potential converts to Christianity. Because of the judicial elites' intimate connections to the plantation owners, including the fact that many judges also owned plantations, they energetically enforced labor discipline and protected property through larceny convictions. Thus, the courts served to buttress the power of economically dominant groups, yet did so under the aegis of moral reform, particularly the reform of uncivilized outsiders.

Yet, despite their interests in supporting the plantation system, judges in this court were careful to weigh evidence. They refused to convict in its absence, even for adultery or failure to work in the case of contract laborers. Contract laborers who complained that they received insufficient food or were forced to do tasks too heavy for their abilities very occasionally won, particularly when they were represented by an attorney such as Nawahi. Thus, the legal system basically supported the groups in power, but it did so through the framework provided by the law. Although the vast majority of contract labor violators were found guilty (97 percent), the legal system represented a screen through which the plantation owners controlled them. It contained the possibility, although remote, of circumscribing the power of the owners over the workers. Moreover, when conflicts in the cane fields turned violent, as they sometimes did, the courts provided a different terrain for workers to contest their subordination. Few did better there than in the fields, but over the generations, they learned to use the legal system as a check on the power of the employers.

The protection offered workers in the courts was far greater when the workers were Native Hawaiians or Portuguese than when they were Chinese or Japanese. In the mid-nineteenth century, there was some effort to protect workers from the abuse of their employers, especially when they were Native Hawaiians. But as the workforce increasingly became immigrants who did not speak Hawaiian and lived in different ways and had recreational lives more unfamiliar to the established elites—as they became more culturally distant from the judges—the judges seemed to feel less paternalistic responsibility. The number not convicted for refusal to work or attacking a boss dropped off. These newcomers were also far less likely to have lawyers representing them. Judicial concern for evidentiary niceties began to fade as the plantations became stronger and the gulfs of culture and class between the court and the defendants widened. At the same time, the local elites from which judges and attorneys were drawn lost their dominant economic position in the plantations as local plantations were absorbed into larger conglomerates based in the U.S. as well as in Hawai'i. Large financial agents, called "factors," controlled shipping and harbor facilities and began to buy up plantations. Nor did local judges feel the same paternalistic concern for court defendants from Asia as they did for Hawaiians who were their Christian converts. The Portuguese fared better than the Asians because of their race and religion. During this period, late in the nineteenth century and early in the twentieth, decisions turned ever more against the immigrants. The Puerto Ricans, arriving at the turn of the century, when the plantation system was fully established, seem particularly poorly treated in court. In some cases, they were found guilty of assaulting their bosses even when the evidence seems ambiguous or unpersuasive. It is clear that the combined features of racism and fear of "others" imagined to be dangerous and undesirable for the social fabric contributed to a diminished concern to weigh evidence carefully.

It is also likely that the local judiciary became less independent as its economic power declined. Indeed, Hapai, the judge of the late nineteenth century, was the only judge in Hilo who did not own a plantation and the only one who was not white. He was also politically weaker, as the protegee of the powerful Severance instead of a politically powerful person in his own right, as were the previous judges. Thus, as economic power was centralized, the autonomy of the local judiciary was diminished, reducing their ability to resist the powerful economic interests supported by the work of the courts.

Doing Ethnography in the Archives

Ethnography in the archives means setting the caseloads of the courts in the context of the people who were running them, and those caught in them, as well as within the context of broader economic and political changes. This contextualization makes it possible to interpret the meaning of the kinds of cases that arrive in the courts and the kinds of defendants who are there, as I have tried to show here. Such a reading provides a way of making sense of the pattern of cases and decisions appearing in this court.

My ethnographic work on the meanings of gender, ethnicity, race, and class in Hilo today provides another way to read back into the past. This study of the contemporary court shows how local social understandings of these identities shape some aspects of legal process in the present, and presumably also did so in the past. In late twentieth-century Hilo, for example, a person's identity depends on his or her history and that of his or her ethnic group. Whites, called *haoles*, are considered a very different ethnic group from the Portuguese, although both would be considered white in other contexts. Members of both groups insist on these differences. The explanation lies their different histories in Hawai'i, since the *haoles* typically served as plantation managers, while the Portuguese were brought to Hawai'i in the late nineteenth century to work as laborers in the field. Although they typically were promoted to supervisory positions and many left the plantations to become crafts-men and skilled workers, the historical distinction between the managerial class and the laboring class produces persisting differences in these identities. For all groups in Hilo, the meanings of identity depend on the history of the plantation and the earlier history of conversion and colonialism of the Native Hawaiian population. More recently, the heroic aspects of the labor movement, lead by Japanese and Filipino workers, along with the heroism of Japanese soldiers (all of these groups are American but label themselves according to their nation of origin) in the Second World War, have provided a new history of identity. Lingering resentments against the haole manager class and its racial supremacy and paternalism continue to shape contemporary ethnic relations.

The court today responds to contemporary concerns, such as domestic violence, according to such categories of identity, forged over time. Moreover, its procedures are shaped now, as they were in the nineteenth century, by ideas about which kinds of people are troublesome or dangerous and which are not. Now, as then, shifting caseloads depend on what is defined as a serious social problem and what is not. It is usually some aspect of the behavior of newcomers, particularly in a conservative, small town environment with a fairly stable elite, which is viewed this way. This pattern appears clearly in the historical record. Problems shifted from opium in the 1880s, gambling in the 1890s, sex with girls under sixteen in the 1940s, cock-fighting in the 1950s through 1980s, to the growing of marijuana and traffic viola-tions in the 1970s through 1980s. There was an explosion of cases of domestic violence in the criminal court in the 1990s: By the late 1990s such cases comprised one-third of the probation caseload. Contemporary ethnography helps us to under-stand the local political and cultural processes that define certain behaviors as crimes and lead them to be brought to the court and prosecuted.

Thus, in order to investigate social change, archival research is essential. Historical data provides clear evidence of changes in the kinds of problems in court and links these changes to shifts in the personnel running the courts and the political currents of the time. The nature of immigration, demands for plantation labor, the expansion of sugar plantations, the pressure for annexation to the United States, and concepts of white racial supremacy were clearly relevant to what happened in the courts. An historical approach is necessary to demonstrate the linkages between court processes and the changing social order.

On the other hand, ethnography is necessary to situate these changes in a local place and with a cast of characters. An ethnographic approach to history unveils everyday behavior rather than only dramatic historical events taking place in capital cities. Much as Foucault argues for attention to the microphysics of power embed-ded in the margins and interstices of institutions, ethnography based on archives such as court records and personal letters provides a way of looking at the everyday exercise of power and resistance. Such an approach contributes to a growing theo-retical interest in the way law constitutes everyday life through the local processing of cases (e.g., Merry 1990; Conley and O'Barr 1990; Starr 1992; Sarat and Kearns

1993; Brigham 1998; Ewick and Silbey 1998). Law shapes social life as it punishes some actions while letting others go, defining some kinds of people as criminal and others as benign. The particular vision of social order and identity shared by those groups in control of the courts is of special importance in this constitutive process.

Appendix

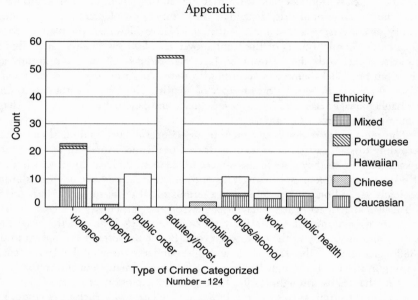

Table 7.1 Ethnicity by Type of Crime, 1853

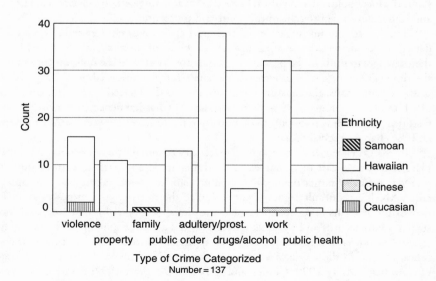

Table 7.2 Ethnicity by Type of Crime, 1863

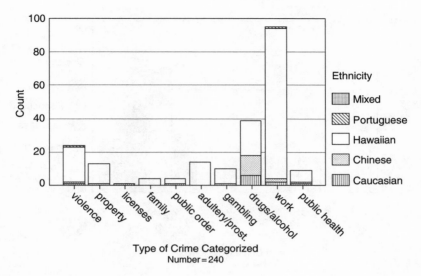

Table 7.3   Ethnicity by Type of Crime, 1873

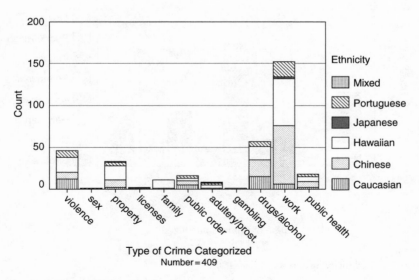

Table 7.4   Ethnicity by Type of Crime, 1883

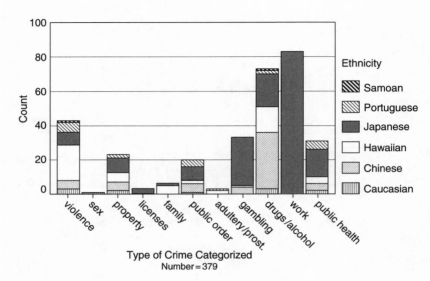

Table 7.5   Ethnicity by Type of Crime, 1893

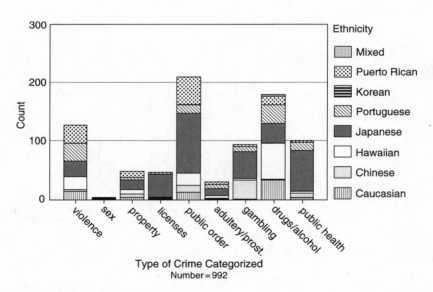

Table 7.6   Ethnicity by Type of Crime, 1903

Notes

Presented at the Law and Society Association Meetings, June 1998. The research described in this chapter was generously supported by two grants from the National Science Foundation (Cultural Anthropology and Law and Social Sciences Programs, #SES-9023397 and #SBR-9320009), the National Endowment for the Humanities, and the Canadian Institute for Advanced Research. I also received support and encouragement from the Hawai'i Judiciary History Center in Honolulu. The research described in this chapter is more fully presented in *Colonizing Hawai'i: The Cultural Power of Law* (Princeton University Press, 2000).

1. These records are housed in the Hawai'i State Archives in Honolulu. The minute books from the district court have been preserved in a virtually complete set from Hilo from the 1850s until 1910, but subsequent records were destroyed. I recorded every case for each sample year: 1853, 1863, 1873, 1883, 1893, and 1903 (see Tables 7.1 through 7.6, this chapter). The work of coding was laboriously and carefully done by several research assistants: Marilyn Brown, Joy Adapon, and Erin Campbell. The District Court was located in Hilo and served only its environs. When the District Court was first established, it was declared a police court because its responsibilities included significant involvement with foreigners.
2. I coded all cases in the Circuit Court in the following years: 1905, 1915, 1925, 1935, 1945, 1955, 1965, 1975, 1985. In the last year, out of a total of 436 cases, I coded only 250. The Circuit Court covered the entire island of Hawai'i but I selected only those cases on the Hilo side of the island for analysis.
3. "The Social Role of the Courts in the Hawaiian Kingdom." National Endowment for the Humanities, Jane Silverman, Principal Investigator; Peter Nelligan, Assistant. This project was done in the early 1980s. I am grateful to Harry Ball and Jane Silverman for help in using this data. This data provided the basis for Nelligan's Ph.D. dissertation (1983).
4. Severance Scrapbook p. 32, "Hilo Happenings" in *Daily Commercial Advertiser* 1890.

References

Conley, John and William O'Barr
    1990   Rules versus Relationships. Chicago: University of Chicago Press.
Brigham, John
    1998   The Constitution of Interests. New York: New York University Press.
Ewick, Patricia and Susan S. Silbey.
    1998   The Commonplace of Law. Chicago: University of Chicago Press.
George, Milton C.
    1948 The Development of Hilo, Hawaii, TH or A Slice through Time at a Place Called Hilo. Ann Arbor, MI: Edwards Letter Shop.
Kame'eleihiwa, Lilikala
    1992   Native Land and Foreign Desires. Honolulu: Bishop Museum Press.
Kelly, Marion, Barry Nakamura, and Dorothy B. Barrere
    1981   Hilo Bay: A Chronological History. Prepared for U.S. Army Engineer District, Honolulu. Typescript.
Leithead, A. Scott
    1974   "Hilo, Hawaii: Its Origins and the Pattern of its Growth, 1778–1990." Unpublished B.A. Honors Thesis, Department of History, University of Hawaii.
Lydgate, J. M.
    1922   "Hilo Fifty Years Ago." Thrmu's Hawaiian Annual. Honolulu: Thomas G. Thrum, pp. 101–108.
Merry, Sally Engle
    1990   Getting Justice and Getting Even: Legal Consciousness among Working-Class Americans. Chicago: University of Chicago Press.

2000   Colonizing Hawai'i: The Cultural Power of Law. Princeton, NJ: Princeton University Press.

Nelligan, P. J.
1983   Social Change and Rape Law in Hawaii. Unpublished Doctoral dissertation, Department of Sociology, University of Hawaii.

Nelligan, Peter J. and Harry V. Ball
1992   "Ethnic Juries in Hawaii: 1825–1900." *Social Process in Hawaii* 34: 113–163.

Osorio, Jon
1996   Unpublished Doctoral Dissertation, Department of History, University of Hawai'i at Manoa.

Sarat, Austin and Thomas R. Kearns
1993   "Beyond the Great Divide: Forms of Legal Scholarship and Everyday Life," Law and Everyday Life. Ann Arbor: University of Michigan Press, pp. 21–63.

Sheldon, John G. (1844–1916)
1909   The Biography of Joseph K. Nawahi. Translated from the Hawaiian by Marvin Puakea Nogelmeier 1988. Hawaiian Historical Society. Typescript.

Silva, Noenoe K.
1997   "Ku'e! Hawaiian Women's Resistance to the Annexation." *Social Process in Hawai'i* 38: 2–16.

Starr, June
1992   Law as Metaphor: From Islamic Courts to the Palace of Justice. Albany, NY: State University of New York Press.

CHAPTER 8

STORIES FROM THE FIELD: COLLECTING
DATA OUTSIDE OVER THERE*

Herbert M. Kritzer

Introduction

In recent years there has been increased attention to the value of qualitative research and the methods for conducting that research in a rigorous manor. We now have available excellent discussions of issues of design (King, Keohane, and Verba 1994; Yin 1994), data collection (Douglas 1985; McCracken 1988; Spradley 1979; Spradley 1980), and analysis (Feldman 1995; Miles and Huberman 1994; Silverman 1993; Strauss 1987). My interest in this chapter is on data collection.

There are three archetypal methods of generating qualitative data: reading (relying on existing texts, typically from archival sources), talking (interviewing), and watching (observation).[1] Historians have been very candid about the issues and dilemmas of their data sources, and historiography provides a wealth of guidance regarding the issues involved in relying upon archival materials (Barzun and Graff 1992; Bloch 1953). Users of field interviews and of observational techniques have devoted less attention to the limitations of the information their methodologies produce.[2] Consequently, I will concentrate my attention on talking versus watching as methods of data generation.

A great deal of sociolegal research (as well as sociological and political science research) relies upon open-ended interviews. These interviews typically focus on a combination of factual and attitudinal issues. Sometimes they are a kind of personal oral history interview, in which respondents are asked to reconstruct past events, with the added element of asking for the reasoning behind specific actions. The duration of the interviews is most often in the range of one to three hours. Interviews can be an efficient way to collect data. For one hour interviews, it can be possible to conduct four or five interviews per day if the respondents are close together geographically and their schedules happen to allow the appropriate scheduling.

Much less research relies upon observation (or observation combined with interviewing[3]). This is not surprising. Observation is extremely time consuming, and much of the time spent doing the observation can be relatively unproductive given that it is not possible to restrict the observation to activities or events relevant to the research, and that much of the observation may involve extremely repetitive activities.[4] Even if time is not a constraint, for many of the phenomena of interest to sociolegal scholars it can be difficult to obtain the access necessary to conduct observation. For example, when Sarat and Felstiner (1995) were beginning to plan their observational study of divorce lawyers and clients, it was not at all clear that they would succeed in obtaining the access that the research required (Chambers 1997: 214–219; see also Danet, Hoffman, and Kermish 1980). In addition, because of the time required, observational methodologies tend to limit the number of data sources in ways that can raise questions about generalizability.

Because of the difficulties of observation, it is not unusual that researchers (perhaps with the important exception of anthropologists, who will find nothing of what I say in what follows the least bit surprising) more often opt to collect data through interviews, attempting to capture through the interview process the kind of information that would be obtained if observation could be carried out. Researchers pay unexpectedly little attention to the limitations of the pure interview methodology, even though at least some of these limitations are well documented (see, for example, Briggs 1986; Converse and Schuman 1974), perhaps because few researchers have the opportunity to compare directly the results obtained with the two methodologies. When such comparisons are possible, the differences can be striking.

An Example

Let me illustrate this with an example from my research on contingency fee legal practice in Wisconsin. The design for this research involved three major components: a structured mail survey of Wisconsin practitioners, observation in three different firms (one month in each firm),[5] and a series of semistructured interviews with contingency fee practitioners around the state. Of interest here are the latter two components. In designing the research, I was very aware of the issues of generalizing from what amounted to three observational case studies (even if those settings were specifically chosen to insure variability). The specific goal of the interviews, which were conducted after the completion of the observation, was to assess the generalizability of the observational data. To do this, I sought to design questions that would provide information that was as comparable as possible to the kind of information I obtained from the observation. Among other things this involved:

- a question asking the respondent to give me a tour of what he or she did the previous day ("Can you walk me through what you did yesterday ... What would I have seen if I had sat in this office from the time you got here until you left for the day? How many different matters did you deal with? How many phone calls did you take or return yesterday? In what ways was yesterday typical and atypical?")
- a question asking the respondent to describe the most recent case closed ("Could you tell me about the most recent case you closed? What type of case was it? How did it come to your office? What valuation range did you see it falling into when you first evaluated it? Who prepared the initial draft of the demand letter/brochure: you, another lawyer, a paralegal? Can you walk me through the negotiating history of the case—when was the first demand made, what was the first offer, etc.?")

- a question asking the respondent to describe what would happen if a potential client contacted his or her office about a particular hypothetical case ("Could you walk me through what would happen if your firm received a call from a potential client who described the following case—who would I talk to first, etc.? [I then described a case that involved a twelve year-old bicycling on a sidewalk who was struck by a car exiting from a commercial driveway]")

Given that I had already conducted the observation when I did the interviews, I knew very clearly the type of information I was seeking. I found that I was unable to get information that was even marginally comparable about the details of events and processes.

One can see the clarity of the problem by looking at an sample of my (edited) observational notes of one day in one of the three firms in which I observed, and one of the better (if not the best) responses I got to the first of the listed questions above.[6] The lawyer whom I observed in the example below, "Alex Stein" (a pseudonym—all names in what follows are pseudonyms), works in a small firm (fewer than five lawyers) that largely (but not entirely) specializes in personal injury cases, including workers' compensation. This firm handles routine, run-of-the-mill type cases, most of which are under $50,000 (no one in the firm has ever handled a case that led to a judgment or settlement of over $500,000, and the two senior lawyers have been in practice for over fifteen years). The day described below occurred toward the end of my time observing in the firm; it was reasonably typical, although there was more in the way of settlement discussions than on most days, and there was less in the way of client intake activities. [To assuage the curiosity of readers, I have included notes on the final outcomes of two of the cases dealt with during the day.]

At 7:00 A.M. Alex Stein called my home to tell me that he wasn't going to his gym this morning and wanted to get into the office early; he asked me if I would like him to pick me up and give me a ride (it's −24° out this morning, and I don't refuse). On the ride in, Alex told me that he had worked on the brief [for a case he is appealing] last night for four hours; he had sat down to do one paragraph and ended up revising one subsection.

When we arrive at the office (about 8:00 A.M.), Alex immediately went to work on the brief. He dictates changes. This takes thirty to forty minutes, during which he declines to take one or two phone calls. He gives the dictation tape to his paralegal/secretary, Jim Allen, at 8:55.

At 9:00 he turns to the mail and telephone messages. In the mail was a police report for the accident of one new client, medical records for another couple of cases (including one that prompts Alex to make an observation about the high charge for copying a bill), and a motion (and accompanying brief) from the opposing lawyer in a non-PI (non-personal injury) case he is handling as co-counsel with another attorney.

In reaction to one of these bills Alex comments on a $55,000 medical bill in a workers' comp case. The medical insurer in that case has agreed to pay Alex a fee (20 percent) if he can recover the medical costs from the workers' comp carrier. In fact, the case was referred to Alex by the medical insurer; it would not have been worth his while to pursue the case just for the benefits that would come to the client. Alex comments on the potential for conflicts of interest between the insurer and the client in cases like this. The client comes first.

At 9:30, Alex calls Bob Strong, a client for whom he had been negotiating a settlement the day before. [This call and most others are handled on the speaker phone.] He gets some additional info on Strong's work situation, work duties, and prior medical treatment. Strong tells Alex that he had no related symptoms prior to the auto accident; his former job (he had been dismissed for absenteeism last month) was on a limited-term basis [meaning that he received no benefits], even though it had lasted three or four years. Alex has Strong describe his job duties; this produces a somewhat confused discussion, but the thrust is that most of the work was light duty such as sorting mail, with occasional assignments (perhaps two to three times a month) that involved moderate to heavy over-the-head lifting (getting this information from Strong is not easy, but Alex does eventually get it). Strong missed virtually no work before surgery. Alex explains the adjuster's concern, "It is just as likely that the client's problem with his neck is related to work as to the accident." Alex observes to Strong, "It's not clear what happened here." Alex tells Strong that the situation is difficult, repeatedly telling him that the issue is whether the surgery is really related to the auto accident, and the difficulty of establishing proof. "This is a very iffy case in terms of litigation. I'm concerned whether this is provable in a court of law." During this conversation Alex discusses with Strong the option of filing for bankruptcy to avoid paying other outstanding debts which would absorb most or all of any settlement Strong might receive.

The call ends at 10:00 A.M. Alex comments "This case is going to be a tough one"; Alex comments that Strong mentioned some leg problems that aren't in the record; Alex's view is that Strong does not have good recall on this. Alex tells me more about the bankruptcy option; under Wisconsin law, one can keep up to $25,000 of an injury settlement. Alex is very skeptical about this case; clearly he doesn't want to take it to litigation. Alex says that if the adjuster makes a low offer and Strong rejects it, Alex will encourage Strong to get another lawyer.

Jim (the paralegal) comes in with the revised brief. Alex spends fifteen minutes or so reviewing it.

Alex's partner comes in with a question about whether there can be multiple independent medical examinations (IMEs) when there are multiple insurance carriers. After responding, Alex discusses a drunk driving case with his partner, wondering about whether the facts of the case might restrict the availability of punitive damages. Alex decides he needs to check the statutes; there is also the question of who is responsible. A few minutes later, he checks the statute and finds a clear answer to his question.

At 10:30, Alex called Carl Hopkins, a client in another case (workers' compensation case), to check in. Hopkins's employer has no work available, given the restrictions on what Hopkins can do, given his injuries. Alex mentions he has another client from the same employer (but at a different location). Hopkins comments that "it's rough work." Hopkins has a question about a recruitment bonus he got for finding another employee: He is supposed to get a bonus of $xx if the new employee stays yy months. The employer is hedging on whether he will pay this bonus because Hopkins is not currently working. Alex tells Hopkins that he is probably entitled to the bonus, but that Alex can't really do anything; he advises Hopkins to go to small claims court if it becomes necessary. Hopkins is concerned about problems he might encounter if he went back to work at the employer. Alex strongly repeats the warning "don't quit." Alex tells Hopkins about a case he had settled earlier in the week in which the employee had quit and about the problems *quitting had created. Alex tells Hopkins "don't do anything without talking to me." He mentions the upcoming IME, and Hopkins asks, "What should I expect?" Alex describes how the company that will do the IME operates, stating that "the doctor is looking for ways to save the company money." Alex tells Hopkins that he needs to emphasize how the injury happened. About the doctor who will do IME, Alex comments "there are worse doctors around ... he's on the conservative side ... I'm not looking for great things from him ... be polite ... show him deference." The call lasted about thirty minutes.*

Alex calls back the adjuster, Stan Davis, in Bob Strong's auto accident case. Alex gives Davis some information that Davis had requested (Strong's work history, medical treatment, work situation). Alex explains that his client does not have health insurance because of his LTE status. Alex emphasizes that he has no independent information, but is simply taking what Strong has said at face value; "If you want to interview him, I'm willing to make him available," Alex offers. Davis mentions that he is looking through his notes. Alex says "there is nothing I am aware of [that explains the injury other than the accident]. I've asked him eyeball to eyeball ... there's nothing I'm aware of that would be an intervening cause." Davis comments that "it's a hard one to swallow ... that the auto accident is the sole cause." Davis is clearly having difficulty coming to a decision about what to offer, and Alex is not doing anything here to push him. Alex is holding back to see what he will say. Davis offers $30,000 "to get rid of this one." Alex is stunned; he was expecting $5,000 or $10,000 as an opening offer (and he would have been happy to get it and get out of the case). Alex does not hesitate in leaping to it "let's probe this one ... if causation was not an issue is my demand of $50,000 appropriate?" Davis concedes that if there were no causation issue, the $50,000 demand would be reasonable, but goes on to say that he still has problem with the case. Alex says, "I appreciate the problem you have but here is my problem." Alex goes on to describe Strong's outstanding $17,000 medical bill, and the problem of getting providers, as opposed to insurers, to take a reduction; Alex goes on to say to Davis "I know where you are coming from." The call ends at 11:05, having lasted twenty-five minutes.

Alex comments to me that Strong needs to seriously consider the bankruptcy option. He then immediately calls Strong, and tells him about the offer, commenting that it is "double what I thought he would offer ... a neck surgery with a good result is worth about $50,000 [if there is no causation problem]. We would have serious problems proving the case [at trial]." Alex gives detailed description of conversation with the adjuster (including adjuster's reluctance to come up with a number). Alex recommends a counter at $42,500 and hopes for a $35,000 settlement. He then goes on to tell Strong that "I think you've got to make some tough decisions." Alex goes through various options mostly related to bankruptcy. He offers to take a fee reduction (rather than one-third of total, he will take one-half of what's left after paying the outstanding medical bills and expenses; this would probably be about $9,000 rather than $12,000 if the settlement is $35,000). Alex comments to Strong that he wants to be sure that he (Strong) gets at least as much as Alex gets. Alex goes on to brag a bit: "I've done a fine piece of work on this case. I handled the adjuster the right way." Alex again talks about the bankruptcy option, which would avoid payment to the medical providers, commenting that "I hate to write off the doctor's bill because he made the case for you ... but business is business and the doctor probably makes at least $400,000 a year." Alex asks Strong about other debts. He then suggests some other options, such as talking to the medical providers to see if they would reduce their bills to *avoid getting nothing at all if Strong filed bankruptcy. Alex asks for authority to demand $42,500 and to settle for whatever he can get, and Strong grants him this authority. He again emphasizes the problems with the case, commenting that "it's not a good risk at all." Alex mentions the downside of bankruptcy. The call lasted about fifteen minutes. Alex tells me that he will wait until Monday to call Davis back because he wants Davis to think that Strong really had to think about the offer and that perhaps that Strong had some reluctance. I ask Alex how much time he has spent on this case;[7] he estimates twenty hours, but reminds me that ninety minutes ago he was ready to walk away from it.[8]*

At 11:40 Alex receives a call from a chiropractor (a couple of days earlier Alex had a rather heated conversation with the receptionist at the chiropractor's office). Alex describes what happened. He had called the office wanting to insure that medical records pertaining to his client were sent to him before being sent to a third party insurer in order to be sure there were no errors in the records (he describes the example of an error he recently discovered in the records of another medical provider). Overall the conversation with the chiropractor goes very well; the chiropractor has no problem sending records to Alex for review before sending them to the insurer, but will not erase records, only note corrections. That is fine with Alex. The chiropractor says that he normally sends a draft of his report to the lawyer before doing the final version; Alex is delighted. Overall, it is a very good conversation. Alex comments, "wasn't he great."[9] The call lasts twelve minutes. Alex tells Jim to call the client to let her know that things are smoothed over, and that her husband doesn't need to straighten out the situation with the chiropractor.

At 11:55 another client, Sue Edwards, calls. Alex had been waiting for the doctor's report, which finally has come in. Alex tells Edwards that on first glance the report looks good. Edwards has a notice from DILHR (the agency responsible for workers' compensation); Alex tells her to send the notice to him and he will deal with it. They discuss the case a bit; Edwards is concerned that her postinjury wage potential is reduced. Alex explains that because she is back at work at her former employer that he can't do anything about the larger labor market, even though Edwards is concerned about what might happen if the employer were to close up the local operation. The call lasts eight minutes.

For about fifteen minutes, Alex works on miscellaneous tasks such as looking at his mail, sorting through and balancing a checking account, etc.

At 12:22, adjuster Bob Fox calls regarding a claim. Alex had demanded $75,000; Fox had previously acknowledged that the case was in the $50,000–$75,000 range. Fox offers $62,500 on a take-it-or-leave-it basis, saying that he will not budge. When Alex indicates that he "would like to counter it," Fox essentially says that he would be wasting his time. Alex talks about the problems of getting the subrogation claims reduced, but he gets no encouragement from Fox. This is an underinsured motorist claim, so if they can't agree on a settlement it will go to arbitration; Alex raises that possibility, but Fox says that going to arbitration is no problem as far as he is concerned. The call ends at 12:27. Alex tells me that the "offer is imminently fair" (there were two accidents and there is some uncertainty about which accident caused what). Alex thinks that Fox might be serious that $62,500 is all there is (i.e., take it or leave it), but Alex says he will test it.[10]

From 12:30 to 1:35 Alex leaves the office for lunch.

When he gets back at 1:35, his partner is on the phone with an adjuster. The adjuster would like to talk to Alex. It turns out that the adjuster wants status reports on several cases, which Alex provides.

Alex had tried calling Sarah James, a potential client involved in an accident caused by a drunk driver, several times before lunch, and tries again now; the line was busy earlier, and is still busy.

Alex has a call from another lawyer on the "lawyer to lawyer hotline." Alex is listed as knowledgeable on a particular area. The conversation lasts about ten minutes.

Alex looks at the opposing brief in the non-PI case for which he is serving as co-counsel. The defendant has moved for dismissal on several bases. Alex does not think the other side has a strong argument.

At 2:00, Alex calls another lawyer to discuss the brief he had worked on the previous evening. The lawyer has little to say "I think it [the second argument] is splendid ... I think it's a darn good job." He has no substantive comments. The call ends at 2:10.

Alex returns a call from the Larry Gavin, the other driver in a case he is handling as a UIM claim. Gavin wants to know what's happening. He specifically asks about the injuries suffered by Alex's client, information which Alex does not want to provide. Gavin asks for the name of Alex's client's insurer, which Alex is reluctant to give out because he is concerned Gavin might make a claim against his client (Gavin ran a stop sign, but Alex's client might have been speeding). It turns out that Gavin needs to get documents from the insurer in order to get his drivers license back under Wisconsin's financial responsibility law. Alex explains that his client cannot sign a release because that would jeopardize the client's UIM claim. Alex agrees to call his client's insurer to tell them that Gavin wants to get in touch, and then the insurer can contact Gavin.

Alex and his partner are talking about letters from doctors. Alex notes that a doctor he met with a few days earlier charged $87.50 for a ten-minute conference. His partner tells of a $187 bill from a doctor for a fifteen-minute conference. Neither is really complaining, because in these cases the doctors' letters were central for making the case.

At 2:30 Alex finally gets through to Sarah James. Alex tries to explain something about one of the potential issues in the case; James gets confused and thinks Alex is telling her that no damages are available. Alex clarifies. James had expected to be downtown in connection with the arraignment of the drunk driver, but it turns out she doesn't need to come. Alex had suggested that would be a good time to meet (and he hoped to get a signed retainer at that time). Alex offers to come out to James's home (which is just outside of town). James says that if on Monday the doctor says she can get out, she would just as soon come downtown. They leave it that they will talk on Monday or Tuesday, after the doctor's appointment. James asks about a crime victimization form that she has filled out and returned; Alex tells her that it raises no problems. The call lasted about twelve minutes.

At 2:48, Alex takes a call from a doctor to whom Alex has written for a report (Alex is not at first sure who the doctor is). Alex is sympathetic, but not sure what to say. Alex coaches him on the "magic language." Alex is careful not to push doctor on the substance of opinion but on how to say it. Alex emphasizes that the magic words are that the accident was "a substantial factor"; doctor says "I think I'm willing to say that." The call ends at 2:52. Alex tells me after the call, "That's a surprise." Alex tells me the background of the case, and that he did not expect to get a favorable opinion from the doctor. He has twenty to twenty-five hours into the case, and it's "just warming up."

Alex tells me that his currently active cases seem to him to be a bit more complex than average. He has lots of routine stuff in the hopper but not at the top. Alex comments on the normally routine nature of workers' compensation claims; for those cases the key is "staying on top of them."

Alex tries calling his co-counsel in the non-PI case; the other lawyer is not in, so Alex leaves a message asking him to call back.

Alex calls another lawyer to talk about the bankruptcy issue confronting his client Bob Strong. The discussion suggests that it might be best for Strong to file bankruptcy before consummating the settlement. Alex says he will try to call Strong immediately and make it a conference call. There is no answer at Strong's number.

At 3:45 Alex gets another call from a lawyer on the lawyer hotline. The other lawyer has a question having to do with claim of an adult child; the case is complicated by the fact that the accident happened out of state. Alex is clearly thinking stuff through as he goes. Alex talks about some cases he's had involving recovery by adult children in wrongful death claims. The call lasts about thirteen minutes.

Alex makes notes on his extensive "things to do" list.

Jim brings in the completed brief, which is ready for printing.

Alex reviews the status of medical bills in a couple of cases, and then cleans off his desk (it's about 4:30). [This is a bit earlier than his usual 5:30 departure, but he arrived at the office about an hour earlier than usual.]

He goes over to EconoPrint to drop off the brief to be printed.

While this day's activities differed in some ways from most of the other days I spent in the firm, the overall pattern is reasonably typical. In the course of the day, Alex Stein dealt with at least twenty different cases, often moving quickly from one matter to another (for Stein to keep anything resembling accurate time records would have been virtually impossible). A very large portion of the day was spent on the telephone with incoming calls often disrupting the activity that Stein was working on; Stein had conversations with clients, potential clients, adjusters, medical providers, and other lawyers. No clients actually came into the office (most clients came in only for the intake interview and then to receive the settlement check, with most contact taking place via telephone).

How well can the image captured by the observation be recreated using an interview format? I attempted to do this by using a standard type of question described in the literature on ethnography: what Spradley (1979) calls a "grand tour question." As noted above, this is probably the best response I was able to get to my question and its follow-ups.

Q: I spent the last few months literally following lawyers around. If I had spent yesterday following you, that is, if I had spent the day with you as a "ride along," can you give me a tour of what I would have seen in the course of the day?

A: Let me just refresh my memory by looking at my calendar of yesterday. I do recall several things that did take place. I have a fairly substantial case that I am trying to negotiate a settlement on right now. I set up a conference with another attorney involved in the case. I talked to my client. There are several subrogated carriers involved and I am trying to work out a reduction in the amount that they are asking in subrogation. I was successful in getting all of them to take at least a 50 percent reduction, but I want a 60 percent reduction. Talking to them and then getting back to my client took up several hours of the day yesterday.

Q: How many calls did that involve, do you have a sense? A dozen?

A: Six, seven, eight—somewhere in that range. It was several calls . . .

Q: Did you settle that case yesterday?

A: Not quite. We are awfully close now. We are in the range now that I think it can be done, and we will work that one out. And there were two other calls, two other claims representatives . . .

Q: Different cases.

A: Different cases. I am working on settlements there. I recall sitting down with my associate yesterday and reviewing two PI [personal injury] cases that he has. I was trying to give him an idea of what would be a reasonable demand range, a reasonable settlement range, and a reasonable jury range if we were to go to trial on those cases. They are coming up to trial rather quickly. In addition, I did some things on a couple of worker's comp. cases. I worked on compromise agreements for them, and got one compromise agreement done. I am preparing the actual formal compromise agreement for that case. On another one I prepared a proposal to try and work toward a settlement on a compromise agreement.

Q: On a worker's comp.?

A: On a worker's comp. as well. And another case I got out the request for attorney's fees from the Social Security Administration.

Q: Did you have any calls from new clients yesterday? Potential new clients?

A: In the personal injury field?

Q: Or worker's comp. or Social Security, any of the contingent fee fields?

A: None of those three.

Q: What other kinds of matters did you work on yesterday;? You said you did some family cases yesterday.

A: Yes, because I had a hearing today that took the morning, so I was doing a lot of work on that. And I had some people come in; we are working on a final Marital Settlement Agreement to dispose of some remaining issues. I took care of that, and worked on several Marital Settlement Agreements; actually I did three of those yesterday.

Q: Three different cases?

A: Three different cases.

Q: How many different matters would you say, total, both contingent fee and otherwise, did you touch yesterday?

A: Different number of cases?

Q: Yes.

A: At least fifteen.

Q: Fairly typical?

A: Pretty typical. I usually work a ten or eleven hour day. I am usually here at 7:00 A.M. and I leave at 6:00, and I don't take lunch.

Again, I should emphasize that this is one of the better answers I was able to obtain. Perhaps that was in part due to the respondent's referring to his calendar as a means of refreshing his memory. Yet, while the interview response is not inconsistent with the kind of flow reflected in the field notes, it gives none of the sense of continuity and discontinuity that comes out of the observation. It is highly likely that more extensive probing could have provided more detail from the respondent, but I doubt whether that would have moved the information much closer to the kind of depth obtained from the observation.

A good example of what almost certainly would not have been obtainable was information of the type that came from Alex's interaction with Bob Strong and the adjuster in Strong's case, Stan Davis. Recall that Alex first spoke with Strong and talked about the problems with the case ("this is a very iffy case in terms of litigation"). About an hour later he talked to Davis, hoping to get an offer in the $5,000 to $10,000 range, and was stunned by the offer of $30,000 "to get rid of this one," in the words of the adjuster. He then called Strong back to report the results of the conversation with Davis, during which he patted himself on the back for a job well done ("I've done a fine piece of work on this case. I handled the Adjuster the right way"). While an interview might have provided the information about the higher-than-expected offer, it is unlikely that it would have revealed either the depth of Alex's concern about the case or Alex's bragging about his success in the case. These are the subtle things that will only be picked up through observation.

Implications

That observation will render more detail than an interview seems virtually self-evident. There are, of course, some things that interviews may tap into more directly than will observation, particularly things related to expressions of motivation. However, for understanding the nature of a social or political or legal process, ultimately, nothing is going to replace actually seeing the process in operation. But beyond the accuracy of description, do the differences between interview data and observational really matter?

The best answer to this question must come from a comparison of analyses of data based exclusively on interviews, and data based on observation (usually supplemented by interviews with those observed). To do this, one needs matching pairs of studies, using the two different methodologies that look at more or less the same phenomenon. I have identified three such pairs of studies, all of which involve lawyer/client relationships:

- *Divorce:* A number of researchers have studied the relationship between lawyers and clients in the context of divorce. Drawing on their extensive observation of divorce lawyers interacting with their clients, Felstiner and Sarat find a great deal of ambiguity in the power relationship, with issues of dominance and control constantly in flux and subject to implicit renegotiation (1992: 1495–1498; see also, Sarat and Felstiner 1995). In contrast, Mather, Maiman, and McEwen, who conducted extensive interviews with divorce lawyers, found that the lawyers reported that they largely controlled the direction of their cases, and described the best way to handle them. The researchers found that lawyers report trying to avoid taking cases in which the client will insist on things that they lawyer views as unrealistic or undesirable. The researchers use the metaphor of passenger and driver, arguing that the driver (the lawyer) largely determines both the destination and the route, with the passenger (the client) at best being allowed to do a little backseat driving (Mather, Maiman, and McEwen 1995).

- *Corporate Practice:* Drawing on eighteen months of ethnographic field work in an eighty-lawyer corporate law firm in Chicago, Flood finds that corporate lawyers vary in their degree of responsiveness to their clients (i.e., how willing they are to simply execute their clients' instructions). The lawyers are most responsive to those clients who are seen as having substantial long-term fee potential; a key element of this finding is the variation among clients, even for a corporate law firm (Flood 1987: 386–390). In contrast, Nelson finds that lawyers in the four corporate law firms (also in Chicago) in which he conducted extensive interviews report their clients largely call the shots; there is no sense of any type of systematic variation in how the lawyers in the four firms studied differentiated among clients (see also Heinz 1983, 1988: 250–259).[11]

- *Personal Injury Plaintiffs' (Contingency Fee) Practice:* Drawing upon my observational work, I found that contingency fee practitioners are attentive to the interests of their clients, in significant part because the lawyers rely heavily on satisfied clients as sources of future clients (Kritzer 1998). In contrast, two interview-based studies, Hunting and Neuwirth (1962: 107–109) and Rosenthal (1974: 95–116), found that personal injury clients had little say in the settlement of their claims, although Rosenthal found that lawyers were more responsive to active clients.[12]

In all three of the pairs of studies the interview-based research presented a clear-cut, relatively unambiguous image of lawyer/client control in the settings studied. The images involved can be readily explained in theoretical terms, which leaves the authors confident about their analyses. The researchers do not qualify their conclusions or discuss what factors account for significant variation in the lawyer/client relationship. In contrast, the observational studies present more nuanced images, which tend to contradict the more straightforward results of the interview-based analyses. It is possible that these differences reflect actual differences in patterns in the behaviors studied; I have no way of ruling out this possibility.

However, my experience doing both the observation and the interviews lead me to the conclusion that the differences arise, at least in part, from the nature of the data each method produces. Reviewing the kinds of questions I asked and the answers I obtained, in relation to client control, I easily see that I could have been led to the conclusion that the lawyers have many tools of control and use them in ways that work against their clients' interests. However, the observations made clear the ways in which the lawyers work hard to satisfy their clients. For example, in the case of Bob Strong described in the observational notes, Alex Stein could readily have decided to try to settle it quickly by sharply reducing his demand. Instead, his approach was to keep his higher offer on the table and let the adjuster make the first move. During the observation, the lawyers I spent time with generally pursued this strategy rather than simply looking for the quickest way out of the case (which is what the interview research on personal injury practitioners tends to emphasize). The reason for this is not that the lawyers are all altruistic professionals (although this has some role), but that the lawyers recognize their own long-term self-interest as being served by producing satisfied clients who stay satisfied *and who consequently refer friends and acquaintances in need of legal representation to the lawyers* (Daniels and Martin 1999; Kritzer and Krishnan 1999; Van Hoy 1997); the lawyers do not want clients to go away and later realize that their lawyer sold them short in the settlement process (see Kritzer 1998).

This became clear only during the observation as I watched the lawyers interact with their clients and with potential clients. I saw one lawyer repeatedly spend significant amounts of time talking to persons about cases in which the lawyer had quickly determined he was not interested. This was not productive time for him in the sense of its generating a fee but it was extremely important in attracting future clients. I saw another lawyer take much more time with potential clients referred by prior clients, and even take one case he had doubts about because it had been referred by a prior client; his goal was to do whatever he could to encourage former clients to make referrals and he believed that once a client had made such a referral, paying particular attention to those referrals was a way of encouraging the former client to make still more referrals. Lawyers want to shape the way they are regarded by their clients because that has long-run benefits to the lawyers; for example, Alex Stein's bragging to Bob Strong about how well he had handled the case may have been in part an effort to increase the likelihood that Strong would talk positively about Stein to friends and coworkers who were potential future clients. A third example is one lawyer's "exit" process after a case was resolved: The lawyer would hand the client the check and his business card, saying to the client, "Hopefully, you won't need me again ... If you know anyone who does [need me], please send them in."

It is hard for me to see how I might have designed interview questions that would have turned up these kinds of patterns unless I had anticipated them in advance.[13] In the interviews, the preconceptions of the researcher will determine the questions, and if the literature says that lawyers tend to dominate their contingency fee (or divorce) clients, the questions are likely to be framed in ways that produce responses consistent with that literature. While the observer does not enter the observational setting with his or her mind a *tabla rasa,* what he or she sees will inevitably differ in important ways from initial expectations.[14] While the observer may fail to see some things because of various preconceptions, he or she will usually discover more than is unexpected than will the typical interviewer who imposes more structure on the data collection process.

Others have written on the limitations of interviews. This writing tends to focus on two problems, often related. The first is the general limitations of verbal communication, including

- the difficulty respondents have in understanding questions;
- the difficulty interviewers have in understanding answers; and
- the impact of cultural norms vis-à-vis interpersonal communication more generally (see Briggs 1986).

The second limitation involves the closely related problems of bias and social desirability: Responses may reflect what the respondents want the interviewer to think about the respondent rather than provide the information solicited by the interviewer. Dingwall describes this succinctly: "the products of an interview are the outcome of a socially situated activity where the responses are passed through the role-playing and impression management of both the interviewer and the respondent" (citing Cicourel 1964, 1997: 56; see also Melia 1997: 26).[15]

The point of this chapter is somewhat different: Interviews by their nature get only the information the interviewer solicits (with some relatively rare exceptions), and then usually only in a highly edited or abbreviated form (see Dingwall 1997: 59; see generally Garfinkel 1967). The most skillful interviewers can mitigate this problem to a limited degree, but the difference between what can be seen during observation versus what can be heard during an interview will remain large. Of course, observation does not allow the seeing of what is unseeable, and an interview can bring out to some degree the inner views, thoughts, motivations, and feelings of participants that the observer will typically not "see."

Many types of observational settings *do* allow the watcher to ask question that provide some of that which is unseeable. Beyond this, even when the core of the data is based on observation, systematic interviewing, either semistructured or unstructured, may be integrated with the observation; particularly if such interviewing can be conducted over a series of sessions rather than in a single sitting, it may be possible to probe very deeply, with the interviewing guided by the observation, and future observation guided by interviewing. In reality, most observational studies do combine observation and interview techniques, if for no other reason that researchers must usually obtain some information that falls outside the period of the interviewing (background, prior events, follow-ups regarding things left hanging when the observation concluded, etc.).[16]

The implications of the choice of data collection strategies for sociolegal studies (and other empirical social science research) are obvious. We must constantly be aware of the strengths and weaknesses of the data that we employ. Qualitative researchers often criticize quantitative researchers for relying upon data that are overly structured by the researcher. Yet, the largest portion of qualitative research relies upon interviews that involve very much the same type of structuring. Dingwall puts the problem very neatly when he notes that "interviewers [regardless of whether the interview is structured, semistructured, or unstructured] *construct* data, observers *find* it [emphasis in original]" (1997: 60); "in an interview study, we can pick and choose the messages we hear and that we elicit, [while] in observation we have no choice but to listen to what the world is telling us" (*ibid.:* 64). While this is overidealizing observation—observers regularly fail to see what is right in front of them—observation can be a more "open" form of data collection. While the researcher will not see everything and/or will often see selectively, the nature of the constraints imposed by the researcher differ markedly depending upon the data collection strategy employed.

Notes

*An earlier version of this chapter was presented at the 1998 annual meeting of the Law & Society Association, Aspen, Colorado, June 4–7, 1998. The research reported here was supported by a grant from the Law & Social Science Program of the National Science Foundation (Grant No. SBR-9510976); additional funding was provided by the University of Wisconsin Graduate School and the Department of Political Science's Glenn B. and Cleone Orr Hawkins Trust. J. Mitchell Pickerill, Jayanth Krishnan, and Lisa Nelson provided important research assistance at various phases of the project. Michael Schatzberg and Mark Goodale provided helpful comments, as did those present at the Law & Society Association panel in Aspen. Additional reports from this research can be found on the author's home page at http://polisci.wisc.edu/~kritzer/.

1. Dingwall (1997: 53) describes the three modes as "reading the papers," "asking questions," and "hanging out."
2. See the essays in Miller and Dingwall (1997) for a significant exception to this. Also, there is a large literature in anthropology dealing with the limits of and issues involved in ethnographic fieldwork (see, as one relatively early example, Agar 1980; other work in this genre include Hammersley 1992 and van Maanen 1988).
3. While in this chapter I tend to dichotomize between observation and interviewing, many, perhaps most, observational studies include a significant interview component. Thus, it is probably more accurate to distinguish between studies that rely heavily on observation supplemented by interviews and studies that rely exclusively on interviews.
4. In Dingwall's (1997: 52) words, "Researchers cannot make the field fit their lives, [nor is it] easy to make your life fit the field."
5. Before the research began, I was personally acquainted with one of the lawyers. A second lawyer I met during the preliminary stage of my research. The third lawyer was recommended by someone who knew many people in the local bar (the lawyer was one of several recommended as fitting a set of criteria I felt I needed to insure the variance in settings I was seeking). A fourth lawyer, whom I had met at a state bar function where I presented results from an earlier project, declined to participate.
6. The respondent in the interview was not one of the lawyers I observed.
7. Occasionally I posed questions to the person I was observing; usually my questions had to do with the background of the case, or something very specific, such as how much time had been devoted to the case (an issue that was important to my research, but not something that the lawyers typically thought about in very specific terms).
8. This case settled for $35,000, with Alex taking a fee of $10,000 rather than the one-third that the contingency fee retainer called for; the client did decide to file for bankruptcy.
9. It was fairly common for the lawyer I was observing (picture me in Alex's office, just sitting their on his couch, hour after hour) to comment on what he was doing or on something that had just transpired. This was more true the first few days I was in an office; as I increasingly became part of the furniture, the lawyers felt less need to acknowledge my presence.
10. The case did settle the following week; Alex was able to move the adjuster up only $500 to $63,000. He did succeed in getting the subrogation claim reduced to $11,000, which was a considerably larger reduction than he had expected to be possible.
11. Nelson is quite sensitive to the softness of his information on lawyer–client relations, given that it was based entirely on the reports of his lawyer-respondents.
12. It is worth noting that Rosenthal had originally wanted to do an observational study, but did not succeed in obtaining the needed access (Rosenthal 1974: 179–180); he was able to conduct interviews with both lawyers and clients.
13. Certainly it is possible that some theoretical perspective would have led me to anticipate the specific patterns that I found, but that would have undoubtedly left other kinds of gaps. It is also possible that the standard "clean-up" type question, "Are there any questions I did ask that I should have?" or "Are there topics that you thought I might have covered that I missed during our conversation?" might turn up unanticipated issues or patterns.
14. I do not want to glamorize field observation as a data collection technique. Anthropologists have written extensively on the problems inherent in field observation, and the problems that researchers encounter in what they see and hear; perhaps the best known of these concerns the controversy surrounding Margaret Mead's work in Samoa (Freeman 1983, 1999; Mead 1928).

15. A third related problem is that interview responses may not be accurate, either because the respondent lied or because the respondent is mistaken in either memory or self-perception (Heumann 1990: 201). Of course, there is also the problem that those being observed may behave differently when an observer is present.

16. One type of question that I would frequently ask during the observation, if the information was not otherwise volunteered, was for background on a case the lawyer was working on. I had in fact conducted a very long interview with one of the lawyers I observed during the planning stages of the research; I had not known the lawyer before that time, and the interview created a framework such that I felt comfortable asking him to participate in the observational part of the research. For about a year after I completed the observations, I regularly touched base with the three lawyers to find out if there had been any interesting developments in the cases the lawyers had been working on while I was in their offices.

References

Agar, Michael H.
  1980   The Professional Stranger: An Informal Introduction to Ethnography. New York: Academic Press.
Barzun, Jacques and Henry F. Graff
  1992   The Modern Researcher. Boston: Houghton Mifflin.
Bloch, Marc
  1953   The Historian's Craft. Translated by Peter Putnam. New York: Vintage Books.
Briggs, Charles L.
  1986   Learning How to Ask: A Sociolinguistic Appraisal of the Role of the Interview in Social Science Research. New York: Cambridge University Press.
Chambers, David L.
  1997   "25 Divorce Attorneys and 40 Clients in Two Not So Big but Not So Small Cities in Massachusetts and California: An Appreciation." 22 Law & Social Inquiry 209–230.
Cicourel, A. V.
  1964   Method and Measurement in Sociology. New York: Free Press.
Converse, Jean M. and Howard Schuman
  1974   Conversations At Random: Survey Research As Interviewers See It. New York: John Wiley and Sons.
Danet, Brenda, Kenneth Hoffman, and Nicole Kermish
  1980   "Obstacles to the Study of Lawyer-Client Interaction: The Biography of a Failure." 14 Law & Society Review 904.
Daniels, Stephen and Joanne Martin
  1999   " 'That's 95% of the Game, Just Getting the Case': Markets, Norms, and How Texas Plaintiffs' Lawyers Get Clients." 20 Law and Policy 377–399.
Dingwall, Robert
  1997   "Accounts, Interviews and Observations," in G. Miller and R. Dingwall, eds., Accounts, Interviews and Observations. London: Sage Publications.
Douglas, Jack D.
  1985   Creative Interviewing. Beverly Hills, CA: Sage Publications.
Feldman, Martha S.
  1995   Strategies for Interpreting Qualitative Data. Thousand Oaks, CA: Sage Publications.
Felstiner, William L. F. and Austin Sarat
  1992   "Enactments of Power: Negotiating Reality and Responsibility in Lawyer–Client Interactions." 77 Cornell Law Review 1447–1498.

Flood, John
1987   "Anatomy of Lawyering: An Ethnography of a Corporate Law Firm." Unpublished Ph.D. dissertation, Northwestern University.
Freeman, Derek
1983   Margaret Mead and Samoa: The Making and Unmaking of an Anthropological Myth. Cambridge, MA: Harvard University Press.
Freeman, Derek
1999   The Fateful Hoaxing of Margaret Mead: A Historical Analysis of Her Samoan Research. Boulder, CO: Westview Press.
Garfinkel, H.
1967   Studies in Ethnomethodology. Englewood Cliffs, NJ: Prentice-Hall.
Hammersley, M.
1992   What's Wrong with Ethnography: Methodological Explorations. London: Routledge.
Heinz, John P.
1983   "The Power of Lawyers." 17 Georgia Law Review 891–911.
Heumann, Milton
1990   "Interviewing Trial Judges." 73 Judicature 200–202.
Hunting, Roger Bryand and Gloria S. Neuwirth
1962   Who Sues in New York City? A Study of Automobile Accident Claims. New York: Columbia University Press.
King, Gary, Robert O. Keohane, and Sidney Verba
1994   Designing Social Inquiry: Scientific Inference in Qualitative Research. Princeton, NJ: Princeton University Press.
Kritzer, Herbert M.
1998   "Contingent-Fee Lawyers and Their Clients: Settlement Expectations, Settlement Realities, and Issues of Control in the Lawyer–Client Relationship." 23 Law & Social Inquiry 795-822.
Kritzer, Herbert M. and Jayanth Krishnan
1999   "Lawyers Seeking Clients, Clients Seeking Lawyers: Sources of Contingency Fee Cases and Their Implications for Case Handling." 20 Law and Policy 347–375.
Mather, Lynn, Richard J. Maiman, and Craig A. McEwen
1995   " 'The Passenger Decides on the Destination and I Decide on the Route': Are Divorce Lawyers 'Expensive Cab Drivers?' " 9 International Journal of Law and the Family 286–310.
McCracken, Grant
1988   The Long Interview. Newbury Park, CA: Sage Publications.
Mead, Margaret
1928   Coming of Age in Samoa: A Psychological Study of Primitive Youth for Western Civilization. New York: W. Morrow.
Melia, Kath M.
1997   "Producing 'Plausible Stories': Interviewing Nursing Students," in G. Miller and R. Dingwall, eds., Producing "Plausible Stories": Interviewing Nursing Students. London: Sage Publications.
Miles, Matthew B. and A. Michael Huberman
1994   Qualitative Data Analysis: An Expanded Sourcebook. Thousand Oaks, CA: Sage Publications.
Miller, Gale and Robert Dingwall, eds.
1997   Context & Method in Qualitative Research. London: Sage Publications.
Nelson, Robert L.
1988   Partners with Power: Social Transformation of the Large Law Firm. Berkeley: University of California Press.

Rosenthal, Douglas E.
    1974   Lawyer and Client: Who's in Charge. New York: Russell Sage Foundation.
Sarat, Austin and William L. F. Felstiner
    1995   Divorce Lawyers and Their Clients: Power and Meaning in the Legal Process. New York: Oxford University Press.
Silverman, David
    1993   Interpreting Qualitative Data: Methods for Analysing Talk, Text and Interaction. Thousand Oaks, CA: Sage Publications.
Spradley, James P.
    1979   The Ethnographic Interview. New York: Holt, Rinehart and Winston.
    1980   Participant Observation. New York: Holt, Rinehart and Winston.
Strauss, Anselm L.
    1987   Qualitative Analysis for Social Scientists. New York: Cambridge University Press.
Van Hoy, Jerry
    1997   "Getting Clients: Supply and Demand Among Plaintiff's Personal Injury Attorneys in Indiana." Paper presented at meeting of Law & Society Association, St. Louis, MO, May 26–June 2, 1997.
van Maanen, John
    1988   Tales of the Field. Chicago: University of Chicago Press.
Yin, Robert K.
    1994   Case Study Research: Design and Methods. Thousand Oaks, CA: Sage Publications.

CHAPTER 9
DOING ETHNOGRAPHY: LIVING LAW, LIFE HISTORIES, AND NARRATIVES FROM BOTSWANA

Anne Griffiths

The power of narrative in constituting social relations is one that has been acknowledged by a number of disciplines, including law.[1] Narrative endorses a multiplicity of forms from a diverse range of sources. But regardless of the form they are given, a key issue is always the basis upon which narratives are accorded recognition, or conversely, denied legitimacy. A second focal question asks how they are situated with respect to other narratives and the types of authority that they command.[2] My chapter focuses on narrative, in terms of life histories, from a village in Botswana, in southern Africa. It does so in order to provide an account of law that is based on people's experiences in daily life in that village, Molepolole. This perspective is one that provides a counterpoint to the type of legal analysis that is associated with legal centralism or formalism.

The narrative that derives from a formalist model of law is one that is based on written texts embodied in legislation and judicial decision making that form the heart of conventional legal discourse. It centers on law-as-text through a rigorous exposition of doctrinal analysis founded on a specific set of sources and institutions. As a result, "law" is confined to a particular framework, one that sets it apart from social life and that promotes an image of autonomy that is used to maintain its power and authority over social relations in general, thereby sustaining a notion of hierarchy while at the same time maintaining an image of neutrality and equality within its own domain.[3] These are the key features of narrative that this vision of the law promotes, but because of its limited terms of reference (which are confined to a discussion of a particular set of institutions, personnel, and texts), such a perspective excludes—in overly rigid terms that in practice cannot be uniformly applied—other forms of narrative that do not conform to its requirements. Divergent forms are treated as "other" and denied legitimacy, thus impeding alternative approaches to conflict resolution. But these "other" narratives have challenged and continue to contest this essentialist model of law—whether in its own terms or through other means. Feminists, to take one example, have focused on what such a model of law ignores, namely, the gendered world in which women live, a world in which the reality of power and authority—acquired through a coalescence of economic, social, political, and ideological factors—undercuts the prescribed status of individual equality and neutrality accorded by law. Many feminists argue, as do I, that such a gendered world operates to the detriment of women.[4]

My chapter highlights these "other" narratives through a presentation of life histories that portray village people's perceptions of law, the circumstances under which they do or do not have access to formal legal forums, and, in particular, the conditions under which individuals find themselves silenced or unable to negotiate with others in terms of day-to-day social life. The last is especially important as it is at the level of daily life that the power and authority to negotiate with others has the greatest impact on individuals' lives. As elsewhere, few negotiations extend beyond daily life into disputes that require handling in a formal legal arena, such as a court. For this reason, recent legal scholarship has moved toward studying law as part of the everyday, through the use of individuals' narratives that may be juxtaposed against those of the official legal system.[5] This move represents a more anthropological approach to the study of law through its shift toward an analysis of the specific, concrete, lived experiences that inform people's lives.

Such an approach, however, that adopts an ethnographic perspective on law, is one that is absent from most lawyers' analyses of law. The importance of the ethnographic perspective—grounded in the detailed study of the particular ways in which individuals form part of a social group—is that it countermands one of the major criticisms leveled at studying narrative: that it merely represents an ad hoc account of individual experience that is singular in nature and that cannot, therefore, be read at a metalevel beyond that of the individual concerned. Ethnography, on the other hand, while documenting individuals' experiences, also traces the connections of these persons to the broader social polity to which they belong and details the ways in which they are located within it. My use of narrative, derived from the life histories that provide the focus for this chapter, is one based on an ethnographic approach to the study of law.

Context of Fieldwork

I became involved in this type of study when faced with the task of setting up a foundation course on family law for the newly established law department at the University of Botswana in 1981. It was then that I had to come to grips with a form of law that is referred to as customary law, one that exists beyond the written texts embodied in legislation and judicial decision making. Such law, which is specific to a particular polity, is based upon oral transmission and does not depend upon the written word for its legitimacy. The need to explore the connections between customary[6] and Western-style law, known as common law in Botswana,[7] led me to carry out field research among a local community. This integrated approach to the study of law had then not yet been undertaken in Botswana.[8]

My research was carried out among Bakwena[9] in their central village, Molepolole, between 1982 and 1989.[10] At the heart of Kwena affairs in terms of the social, political, and ritual life of the *morafe* (polity, often referred to as "tribe"), Molepolole also acts as the regional center for Kweneng District. It is only seventy kilometers from the capital city, Gaborone, to which it is connected by a tarmacadam road and by daily bus services. Its population, which was estimated to be 20,500 in 1980[11] has steadily grown to almost 37,000 in 1991.[12] However, this population is one that fluctuates due to seasonal migration to agricultural lands and cattleposts for subsistence purposes. Today these activities represent an essential part of a Kwena family's livelihood, together with migrant labor, which is the major source of cash income for most families in Botswana.[13]

Throughout this period I worked closely with Mr. S. G. Masimega, who acted as my guide and interpreter in translating from one official language, Setswana, into the other, which is English. He was an influential person in the village, having acted as personal secretary to the late Chief Kgari Sechele (1931–1962) and having served on the district council and on important committees such as the Village Development Committee and the Parent Teacher Committee. He is known throughout the village as "Mr. Commonsense" and was in his seventies when I first began my research; he died in 1996. This research involved transcribing records from the Chief's *kgotla* and the Magistrate's court and interviewing local personnel. It also involved sitting in on cases and disputes in local wards and *kgotlas* and talking to *kgotla* members.

My focus in 1982 was on cases and disputes. General discussion with people at all levels of society was used to provide the general background within which such cases or disputes could be contextualized. To do this more thoroughly I went back in 1984 and carried out a study of one social unit in the village called Mosotho *kgotla*. This involved working back from persons associated with the *kgotla* in 1984 to field-notes provided by Issac Schapera, who pioneered ethnographic work among Batswana, setting out family genealogies relating to the founding of the *kgotla* in 1937. I acquired detailed information on how people were related and on their life cycle, including education, employment, childbearing, and marriage. I also carried out detailed life histories on a number of families from the *kgotla*. In addition, I followed up, where possible, in finding out what had happened to parties involved in disputes in 1982. In 1989 I focused on updating the life histories and what had happened to those involved in disputes. In this way I acquired a longitudinal perspective on what had happened to parties in disputes, through an extended case study. I also acquired a picture of continuity and change across two generations, which highlighted the differences between and among the sexes.

The Household and the *Kgotla*

As with any Tswana village, the organization of Molepolole is structured through administrative units, known as *kgotlas* and wards, which derive from households. In Tswana ideology, a household consists of a male head with a wife and their children; if a man has more than one wife, there is a separate household for each. This ideal has long since ceased to be realized. NDP6, a government development plan comments, that "females head a third of the households in urban areas and half in the rural areas."[14] UNICEF has also commented on this in its observations on children, women, and development in Botswana.[15] A subsequent government development plan, NDP7, observes that "women headed 40 percent of households in urban areas and nearly half in the rural areas."[16] Kocken and Uhlenbeck[17] observed that women tended to establish their own households at a later stage in the life cycle than their male counterparts. More recently Alexander[18] noted that male heads of households are on average five years older than female heads in both rural and urban areas and that the average age for both female and male heads is eleven years higher in the rural areas than in the urban areas. The term "female-headed household," which appears in the literature on government planning and policy development in Botswana[19] is one which is the subject of some controversy.[20] I use the term "female head of household" to denote those households associated with Mosotho *kgotla,* in which women are in de facto control of the household and lands attached to it and where no adult male of equivalent generational status, whether husband, partner, or brother, was present at the time of research.

Because of these complex realities, I use the term "household" to represent a physical entity or domain which is located in or associated with a *kgotla.* A ward is composed of a number of households organized by the chief around a *kgotla.* In precolonial times, all households of a ward were at least nominally descended from an eponymous founder, but this has not been the case for several decades. It is through households that the political structure of the morafe maintains itself.

A *kgotla* is the assembly center (both the physical location and the body of members) of a group of households presided over by a male headman or wardhead; in the past, all household heads were related through the male line, but this is rarely the case today. It forms part of the organization of Tswana society that revolves around the construction of a *morafe.* The political community within Kwena society, like that of other Tswana *merafe* (plural of *morafe,* polities), is conceived of as a hierarchy of progressively more inclusive coresidential and administrative groupings, beginning with households, and expanding to cover extended family groups in *kgotlas,* to wards. Wards are the major units of political organization of a Tswana village; they are still presided over by men. The most powerful ward is *Kgosing.* The word *Kgosing* is derived from the word *kgosi,* interpreted as "chief." *Kgosing* refers both to the chief's ward, which is the most senior of all Kwena wards in the polity over which it presides, and to the chief's *kgotla* (often, loosely, called the chief's court), which lies at the heart of *Kgosing* ward, representing the most senior of all the *kgotlas* within the ward itself. I use *Kgosing* to refer to the ward and chief's *kgotla* to refer to the court. The chief's *kgotla* in *Kgosing* is the most senior and powerful in the polity and represents the apex of the administrative and political structure through which the *kgosi* exercises his power. When I began my research in Molepolole in 1982, there were six main wards[21] and seventy-three *kgotlas.*[22]

Life Histories

Originally information was gathered to provide a broader social background for my study of disputes. However, I have come to realize how crucial these life histories are in shaping the context in which people structure and pursue negotiations and claims in daily life, or, alternatively, desist from doing so. The narratives drawn from these detailed life histories also highlight the gendered world in which women and men live and how this affects women's access to law. How individuals are situated in terms of networks concerning kinship, family, and community, and the features that affect their position within material and symbolic hierarchies are crucial. Such elements form part of the multiple ways in which individuals are inscribed as social persons.

As such, they have an impact on the forms of expression or types of discourse that individuals employ and that affect their ability to negotiate statuses as well as to articulate claims with respect to one another. What is key is the role that access to resources (including family histories) plays in constructing different forms of power; this in turn impacts upon individuals' abilities to negotiate with one another. Such power informs the kinds of claims that women and men can make on each other and, given the gendered nature of the world from which such power derives, this creates challenges for women extending to their use of both Tswana customary and common law.

Within these networks marriage plays a central role in delineating members' obligations towards one another and also defines men's and women's social relationships. The life histories gathered from Mosotho *kgotla* demonstrate that women's access to resources, including marriage, is heavily dependent upon the type of network to which they belong. Such networks embody two basic forms of existence that have emerged during the course of the encounter with colonialism and which mark the process of social differentiation that is in operation in Botswana today.[23] These encompass, on the one hand, those whom Parson[24] has termed the "peasantariat," a group whose existence is predicated upon subsistence agriculture, the raising of livestock, and migrant labor of an unskilled nature, and who represent the majority of families in the country. They may be contrasted with those whom Cooper[25] refers to as the "salariat," whose existence is dependent upon education and the acquisition of skilled and secure employment, which among the younger generation is predominantly government based. This group, which represents a national elite, no longer engages in subsistence agriculture or in migrant labor of the type entered into by the peasantariat.

Within Mosotho *kgotla* there are families who span both these groups and who highlight the important consequences that membership within these networks has for individuals, especially women, when it comes to negotiating statuses and rights to property. These are, of course, not rigidly fixed categories—the salariat could not exist without the possibility of social mobility—and should be thought of as the polar regions (not end points) of an uneven continuum of socioeconomic-political statuses. They do, however, reflect quite accurately the lived reality of the vast majority of Batswana (citizens of Botswana) lives, as the authors in Kerven[26] demonstrate for the period of this study. From the life histories culled from over two generations of those living in Mosotho *kgotla,* certain patterns of existence are foregrounded that revolve round networks of varying kinds. The networks to which women belong underline the ways in which power is constituted in the world of the everyday life that is crucial to an individual's existence, as well as highlight the important factors that inform people's actions before disputes arise or parties turn to courts to settle their differences. Despite the demonstration by Starr and Collier[27] that power was now central to legal anthropology, this kind of analysis remains absent from much recent research in the discipline, which has focused instead on disputes as the site for studying conflict and the maintenance of social relationships. My work reasserts and extends that of Starr and Collier.

The different life trajectories open to individuals are exemplified by the descendants of one of Mosotho *kgotla*'s founding ancestors, Koosimile, who had two sons (Radipati and Makokwe) by different wives. Makokwe's descendants, who represent the majority of families in the *kgotla,* engage in subsistence agriculture, the raising of livestock, and migrant labor of an unskilled type associated with the peasantariat. Radipati's descendants, however, have pursued another form of existence, one founded on education and skilled, secure employment, which has placed them among that elite salariat nationwide. Within these groups women find themselves differentially situated from men in terms of the kinds of claims that they can pursue with respect to status and property. Not only that, but their position in relation to one another also varies according to their affiliation within a particular group, as we shall see with three women—Olebeng, Diane, and Goitsemang—who, although of the same generation and related to each other, have had very different lives.

For those families in Botswana who focus on subsistence agriculture, the raising of livestock, and migrant labor of an unskilled nature marriage still play an important role in providing access to the broader networks of suprahousehold management and cooperation on which they rely for their subsistence. This is so for Makokwe's family, in which there has been a relatively high rate of kin marriage among members of the older generation (ranging from fifty to ninety years old). When it comes to marriage within this group, women find their choices shaped by male networks and structures of authority that provide the resources and mainstay for their existence. So, for example, through male sibling support some women find themselves with the power of choice that is not available to other women who lack access to this type of network. This gives rise to a situation in which those with choices, such as Olebeng, may opt not to marry, while those without access to the conditions under which such choice become available—like Diane, whose situation we shall examine in a moment—want to marry but are unlikely to do so because of the position in which they find themselves.

Olebeng: An Unmarried Woman with Supportive Male Siblings

Life for the women in Makokwe's family revolves round the village and the lands where they engage in domestic and agricultural activities. Living in the village for much of the year, they are able to attend school, unlike their brothers who are away herding cattle at distant cattleposts. This means that they are able to acquire a greater degree of formal education than their brothers, who in many cases among the older generation, have received none at all. Among this group, women's work is integrated with that of their male counterparts. Makokwe's only daughter, Olebeng, for example, has been part of a family network exchanging her domestic and agricultural labor for her brothers' assistance with ploughing and support. She has never moved beyond this sphere of operations to undertake any form of paid employment, so that throughout life her activities have been such that they have linked her into a network where she has had to rely on male support from her father, her brothers, and her male partners for existence. Within this system, she has been fortunate because her five older brothers have been quite generous in providing support, assisting her with ploughing and upholding her welfare. They have, for example, consented to her taking over the natal household because they have all married and established their own households elsewhere in the village. Compared with other women, she is in a relatively strong position in that her circumstances have permitted a certain degree of choice, including whether or not to marry.

During her life she has had several children (all of whom died at birth) with a number of male partners but has never married. This is out of choice, according to Olebeng, who maintains that from her very first pregnancy neither she nor her family had any interest in pursuing the issue of marriage.

But Olebeng's situation is unusual because in this environment—where emphasis is placed on subsistence agriculture, livestock, and a cash input from migrant labor—marriage is particularly important for women, as unmarried daughters and sisters find themselves at the bottom of the social hierarchy in terms of power and access to resources. Such power derives not only from status and point in the lifecycle, but also includes an individual's capacity to generate or control resources. Among women, power devolves with age linked to status, so that a young, unmarried, childless woman is in a less influential position compared to her older married sister who has children. Both, however, defer to their mother and even more so to their grandmother, who by virtue of her age and status is considered to be in the most powerful position of them all. It is not age alone but the incidents that mark its passage, such as childbearing, that are integral parts of the lifecycle that create status; the combination of age with status fuels the dynamics of power. The same is true for men. A young childless man who has never experienced formal employment has less status than his older married brother who has children and has worked at the mines. Both should defer to their father and grandfather, who have passed beyond these stages. However, this is not always done, especially where the older generation is dependent on the younger to provide for them through the cash that they remit back to the family from their earnings as migrant laborers.

In these circumstances, women find themselves constrained by the gendered world in which they live, whereas men have a greater degree of control over access to resources, such as cash, on which all households depend for their existence. Money is not only required to support family members, but to maintain the subsistence agricultural base through the purchase of seed for crops, labor (where kin are unavailable or demand payment), and livestock for ploughing. Women have fewer opportunities than men to generate the income required because they do not have access to the most common forms of male employment down the mines or on construction sites—and the returns they receive from participation in the informal economy are insufficient on their own to provide the resources that they require. Those who are most successful are married women whose husbands have paid out the cash necessary to promote and sustain their activities. Women selling fat cakes (fried bread dough), for example, need money to buy the flour with which they are made. Also, while women may have access to land and own livestock, the utilization of these resources is often mediated through men. This is because it is men or young boys who herd the livestock at the cattlepost and who are responsible for moving them to the lands for ploughing or to the village to sell. Not infrequently, a man will report that cattle have gone missing or died. Given the distances involved, it is hard to challenge a herder's claim, even though an owner may suspect that the herder has in fact appropriated them for his own purposes. When it comes to cultivation of land, most people still rely on oxen to work the plough. Those who have a team or can contribute to one have control over ploughing and the sequence it may follow. In Mosotho *kgotla,* these are mostly married men who plough their own fields before those of their brothers, parents, and—lastly—unmarried sisters.

Diane: A Vulnerable Unmarried Female Head of Household

Within this kind of network, an unmarried woman finds herself at a disadvantage. While part of a group formed of her family and kin who have responsibility for her—and thus obligations to plough for her—her position is a vulnerable one in that her interests are subordinated to those of other family members. This is the case with Diane and her unmarried daughters, who have found themselves greatly disadvantaged by the constraints that are inherent in a kinship network. Unlike Olebeng, Diane has had a relationship fraught with conflict with her male kin. Her brothers abandoned her after their father's death. They have not only left her to fend for herself but have also appropriated for their own use the land that she was left by her mother. Without access to her brothers' network she has found herself in a position in which she has had to rely on a series of male partners for support, in relationships of *bony-atsi* (concubinage), that preclude marriage. Despite this, she still firmly expresses the view that "it is natural with Batswana to marry. A woman must marry." Her own life history, however, has placed her in a position where marriage is no longer a viable option given the number of children that she has had with different men.

She describes her first relationship in the following way:

I left school in 1955 when I was seventeen years old because I was pregnant. I met the father at my mother's place in Molepolole. His name was Nelson Guwer and he was from Thosa ward. When I became pregnant my parents took action immediately. They went to his family and discussed marriage. We were both asked if we agreed to marriage and said yes. His parents paid *tlhtlhagora* [compensation for pregnancy prior to marriage] and gave me a ring. At that time he was working in Molepolole. After my confinement he went to work elsewhere and he ceased to visit me. My mother went to see his family and when questioned about marriage arrangements they said they were no longer in a position to marry because my father's elder brother Rampole was not in favor of the marriage. That was not true. That was just a lie. I gave up at that stage. I felt the promise of marriage would never be fulfilled.

After that I met another man who proposed marriage. He was from Mosarwa *kgotla*. His father came to tell my parents that he had seduced me and that he was requesting marriage. They agreed to marriage. We had three children together. Before that third child was born my mother went to see his family. They said that he would marry me when he came back from the South African mines. He never came back and we gave up. He never did anything. He did not support the children. My family supported me. My father was working on the South African Railways and paid for the children's school fees, food, and clothing.

I met another man from Molepolole and had a child with him. He never did anything. Then I met a man from Koodisa ward and had five children with him. He is married. Until he was laid off from the South African mines [in 1983] he supported all the children. We were together over ten years but now he just drops by from time to time to say hello. As he has been without a job for a long while he no longer provides any support.

By 1989 Diane had had ten children by four different fathers. As one of the poorest female heads of a household in Mosotho *kgotla*, her life history fits the national profile of the vulnerable female-headed household described by Kerven,[28] Izzard,[29] Brown,[30] and others which is so much at risk and with which the government of Botswana is so concerned.[31] This is because of the highly impoverished position in which many female heads of households find themselves due to the difficulties they face in acquiring resources without male affiliation or support.[32] This situation is often perpetuated into succeeding generations, because children who have experienced deprivation continue to find themselves disadvantaged as adults in ways which affect their ability to establish and maintain their own households.[33] This is the case with Diane's five eldest daughters, who have all, like their mother, had to leave school early because of pregnancy in circumstances that have not led to ongoing relationships with the fathers of their children. This has continued to be the case when subsequent pregnancies have ensued. Of the seven relationships entered into by Diane's daughters between the ages of 16 and 29, only one has displayed any of the signs associated with a potential customary marriage, but it was very rapidly rejected by the man's family. In these circumstances, given her own situation, Diane found that she lacked power and authority to enter into negotiations over her daughters' pregnancies with the men or their families.

Goitsemang: An Unmarried Woman with Access to Resources

Other women, however, who form part of an emerging salariat find themselves with a greater degree of power and control over the choices that are open to them. This is because within their family group, they are less reliant on the type of male networks that peasantariat women depend on for their existence. Goitsemang is just such a woman. Despite the fact that her father Radipati was Makokwe's half brother, members of Radipati's family have experienced very different life trajectories from that of Makokwe. Unlike his contemporaries, Radipati (who died in 1950) was an educated man who placed great emphasis on his children's education, which his wife Mhudi struggled to provide after his death. As a result, his three daughters were educated (at a time when many women only received a nominal education) and acquired formal employment. The eldest unmarried daughter, Goitsemang (aged fifty-two in 1989) worked as a nurse in South Africa and then in a management capacity for a construction company in Botswana, so that she has been able to build a house in Gaborone. This is something that many people in the village aspire to but are unable to achieve. Her younger unmarried sister has also acquired a plot of land in Gaborone by working for the same company. Radipati's sons were also educated and two of them, most unusual for that time, went on to acquire university degrees. Through their access to education and skilled, stable employment the family fits the kind of profile associated with the emerging salariat. Among the younger generation a number of women are employed as teachers and court clerks, and the men are similarly situated within government employ. The family's activities differ from those associated with a subsistence agricultural base and they no longer plough.

Within this family group, Goitsemang, like her contemporaries Diane and Olebeng, has had children and remained unmarried. However, in her case, her relationships with men had the hallmarks of a potential customary marriage that failed to materialize. Such a marriage reflects a process that takes place over many years and involves reciprocal relations between the respective families. Unlike a civil or religious marriage that is registered, it is not necessarily predicated upon a specific, identifiable occasion. While some features, such as the transfer of *bogadi* (marriage payment), may be treated as definitive markers of marriage, their absence does not rule out social recognition of a relationship as a marriage. Among Bakwena, a ceremony, *patlo*,[34] is viewed as being central to the constitution of a customary marriage. But it is not the only definitive feature, so that relationships can and do acquire the status of a marriage without it. What is important is the degree to which both families have become involved in the relationship and accorded it public recognition through the giving of gifts and attendance at significant life events, such as celebrations for the birth of a child, or funerals.

Goitsemang observed that her first partner

> was Charles Magogwe from Maribana *kgotla*. We met in Johannesburg where I was working as a nurse. He promised to marry me. We had a child [a daughter] called Calamity who died and then Charles' parents agreed to our marriage. We had another girl who died in confinement. Charles' parents came to the first funeral but they never gave anything for the second child or came to her funeral.

In this case, the families met, discussed and agreed to marriage on a number of occasions. The man's family provided support for both mother and child and attended the first child's funeral. But by the time the second child was born the man and his family appeared to lose interest, which was demonstrated by the fact that they did nothing to support this child and did not attend her funeral, a very public sign of their dissociation from the relationship. This relationship provides a typical example of the kind of marital negotiations referred to by Comaroff and Roberts,[35] in which the parties start out seriously exploring the potential of the relationship, but over the course of time one of them withdraws from the process. This is often due to the fact that one party has opted to pursue another relationship instead.

In her subsequent relationship, Goitsemang found herself at odds with her family. Her mother Mhudi explained that:

> After the death of the second child she stayed with us here in Molepolole and the family supported her. Her sister Salalenna was working then in Gaborone and could help her. She went to stay with Salalenna in Gaborone and found a job as a clerk. She then met Patrick Kgosidintsi, a businessman from Molepolole. He wanted to marry her and came to see us as parents to propose marriage. As he was already married we told him that we are Christians and said that we could not accept a married man marrying our daughter as a second wife. She was in love with him against our will. He visited her here at home without our consent. She had a child. He brought foodstuffs in a truck but we told him to go away and take the foodstuffs with him as we did not want him to marry our daughter. She went back to Gaborone to work and had another child with him.

This relationship also had the potential for a customary marriage but was rejected by Goitsemang's family because they did not want her to enter into a polygamous union.[36] They were anxious to avoid public recognition of the relationship. That was why they rejected the foodstuffs and all forms of support. Nonetheless, Goitsemang maintained her relationship with Patrick, who is closely related to the Kwena ruling elite,[37] against her family's wishes. She had the power to ignore them because her employment gave her access to a world in which she was beyond her family's control in terms of apportioning resources. What was crucial for Goitsemang was that she was removed from the kind of pressures that accompany dependence on domestic and agricultural labor, and the networks that sustain them. Through her education and training she had access to alternative means of support. These resources empowered her to make decisions on her own account and made her less vulnerable to demands made by kin. Indeed, Goitsemang has felt able to challenge her brother David's claims to control over the natal household under customary law and has received sufficient support from *kgotla* members to continue running the household for the time being.[38] Her situation is very different from that of Diane, who is unable to contest her brothers' actions, or from that of Olebeng, who survives on the basis of a cooperative relationship with her brothers. At her stage in life, Goitsemang has no desire to marry, because, as she observes, "marriage just brings quarrels." This is a view shared by a number of educated and employed women, who prefer to avoid the status of "wife," given the ways in which gender impacts spousal roles.[39] Goitsemang's two daughters, however, are not among this group. Both Eva, who is a teacher, and Patricia, who is a court clerk, have married and live with their husbands in Gaborone.[40]

The Social and Legal Dimensions of Procreation
These life histories and narratives underline the extent to which procreation is an integral part of social life and how it is accommodated within people's lives. For most women, pregnancy forms a natural part of their life cycle, often occurring at an early age.[41] How it is handled varies, but it is clear is that where possible families, not just individuals, enter into some form of negotiation over the conditions in which its social cycle takes place. Makokwe's youngest son, Ramojaki, speaks for many in the *kgotla* when he notes that with his daughter's first pregnancy "we went to his [the man's] family with a view to compensation or marriage." Among Bakwena, compensation is set at eight head of cattle or a monetary sum of 640 pula (about $900 in 1982). In Mosotho *kgotla,* in well over half the cases in which pregnancy occurred, steps were taken by the families to discuss the situation.[42] As noted earlier, such negotiations are essential for initiating and pursuing the processes that may lead to a customary marriage. Where families in the *kgotla* met for discussion, over half deliberated on and agreed to marriage.[43] But only a few marriages have materialized[44] and in the majority of cases the relationships have ended.[45] In some cases an attempt was made to pursue the preliminaries to a marriage but was later abandoned, while in others it was clear from the start that an agreement to marry by one party or the other was just a strategy to avoid paying compensation and was never really under serious consideration. On the other hand, there were several cases in which compensation and not marriage was the focus of discussion,[46] and most of these involved first relationships.[47] But there were also situations in which there was not even the semblance of an agreement because the man and his family avoided meetings or denied paternity. This was Setswamosimeng's experience when her daughter Flora had a child with a boy she met at school in Moshupa. "We wrote to the boy's father and he replied but dodged meetings. I tried to see the boy's parents several times with my sister Mosarwa's eldest son, Gaethuse, but they kept saying that they were at Orapa. They were just avoiding the issue."

The matter was not pursued because "the boy's father was a crook and [I] was afraid that if the matter was pushed regarding marriage he might hurt my daughter. I did not trust him over marriage or compensation." As an elderly widow, Setswamosimeng had to rely on her sister's son to act as the family's male representative in these matters. With no sons of her own, living at a subsistence level and unable to plough for lack of resources and livestock, she is one of the impoverished female heads of households in Molepolole. In these circumstances, like Diane and her children, neither Setswamosimeng nor her daughter had any power to negotiate whatsoever.

In those cases where there was no contact between families this was usually for good reason. In over one-third of the cases,[48] this was because the families knew that they had no claim on the man. Among Bakwena, compensation is only payable for the first pregnancy and cannot be claimed where a second or subsequent pregnancy arises. Baikgodisi, who is in her late forties, observed that her family took no action with her second partner "because I already had too many children with the first man." By that time she had already had eight children. The same was true for Mmanchibidu, whose mother stated that the family did nothing about her second liaison "because she had a child with a man years before."

It is clear that in all these situations individuals and their families assess their options with regard to the social framework in which they live, and in the light of their power to pursue varying forms of negotiation. In this process there are those who find themselves silenced or marginalized and who, as a result, never contemplate the possibility of turning to a formal legal arena. There are also those who may consider such an action only to reject it on the grounds that it will impact negatively on their social relationships, which they cannot afford to put at risk. Although men have reasons to make complimentary calculations, women, whether unmarried or married, have more urgent need to factor in whether the potential benefits of negotiating will be outweighed by loss of respect and/or support from kin. Once again, this requires individuals to make judgments about whether or not they can mobilize the resources necessary to support their position. Thus, all the factors that constrain women in social life also operate to constrain their access to and use of the formal legal system.

Knowledge, Access, and Options

Up to this point, I have shown how the life histories and narratives have highlighted what is at stake for individuals and their families. In some cases, parties happily reach an agreement regarding compensation or marriage. In others, they do not, but parties may not feel obliged to pursue the matter as a dispute. For many the end is neither consensus nor conflict, but often an indeterminate state in which the matter is left hanging or silently slips into abeyance while individuals and their families turn elsewhere to explore another relationship's potential. Nevertheless, these processes may serve as the basis on which a dispute may develop, when one side is dissatisfied and wishes to take the matter further. This happens when negotiations move beyond the families concerned into a more public arena, engaging the participation of a third party who is a public figure. In Molepolole, such a figure may be the headman of a family's *kgotla*, the wardhead or an official associated with the Chief's *kgotla*, or the local Magistrate's court. Ideally, the dispute should pass through the hierarchy of the *kgotla* system before it reaches the Chief's *kgotla* or Magistrate's court; accordingly, it has not been—and strictly-speaking is still not—considered appropriate to raise the dispute directly in the Magistrate's court without first having attempted some form of family or *kgotla* mediation, although now it is increasingly being done.

Women, thus, can and do go straight to the Magistrate's court, for a number of reasons. Some do so because they feel that they will not receive a fair hearing in the *kgotla* due to the man's status and local connections, which will operate in his favor to place him in a more powerful position than the woman. But nepotism is not the only reason why women bypass the *kgotla* in favor of the Magistrate's court. There may be pragmatic considerations, such as avoiding the delay that ensues in the *kgotla* due to the fact that proceedings cannot take place until all the relevant family members are present. Assembling all such members is time consuming, particularly when they are living and working in other parts of Botswana or even outside the country. Or women may simply wish to deal with the matter themselves, free from intervention by family or kin. This is especially the case where their relationship with the man is a tenuous one and there has been little family involvement.

In deciding whether or not to pursue negotiations over pregnancy, many people are aware of the options that are open to them and factor this into their decision making, not only with regard to familial discussions, but also with respect to pursuing a dispute in a legal forum. Knowledge, however, may preclude action. So, for example, Diane "knew after the first child [she] could not use the customary court." She was unable to use the Magistrate's court because "at that time there was no provision for support" (i.e., there was no statute for maintenance, such as the Affiliation Act, in force). She acknowledged that even if such a provision had been in force, she would not have made use of it because "[she] felt it was useless to do anything when a man turned away from his promises." This attitude is not uncommon. The woman's family may act initially to test a liaison's potential, but on discovering the other party's lack of commitment, may let the matter drop.

Some women, such as Seiphimolo (who is one of the few women making a living from running a shebeen[49]) have no interest in pursuing the man's family. Although she knew that she could go to court, she never considered taking legal action because, as she said "my father, Rampole, and my brothers helped me with the children's school fees when they could. I knew that I could go to the Magistrate but did not because I was managing. I was quite able to support the children." She commented that "if I had had a problem with support I would have gone to the Magistrate." She knew the Chief's *kgotla* would not pay compensation for the second child and observed that "even with the first child they like to refer the case back to parents to settle."

Younger women, however, may lack this type of knowledge especially when a first pregnancy is involved. As Oailse Timpa explained, she did not act on her first pregnancy because "I was still young then [nineteen or twenty] and did not know that I could go to the D.C. (District Commissioner) [meaning Magistrate's court].[50] I didn't even know that nursing a child was troublesome." However, Oailse eventually learned from other friends' experiences that she could go to the Magistrate's court, and said that with her second pregnancy "I had by then realized the problems of nursing a baby. A baby needs washing, feeding, and clothing and friends advised me to go. I am very discouraged because every time I go to the D.C.'s office I am asked why I had a child and why I worry about it." Many women learn about the Magistrate's court through other women, especially from friends or relatives. In some cases, their attention is directed towards the court by employers who advise them to raise maintenance claims, or by *kgotla* officials who, in turning them away from the *kgotla*, advise them to pursue their action for support in the Magistrate's court.

In making a decision over how to handle a pregnancy, parties attitudes and actions are to some extent shaped by the activities that inform their lives. Given Radipati's family profile—high levels of education, with women as well as men in employment and linked into the salariat—it is not surprising that Goitsemang's younger sisters both turned to the Magistrate's court. In one case, no action was envisaged while marriage was being contemplated, but when this option was rejected she went to the Magistrate for maintenance "because it was easier to raise the matter before the Magistrate where I was living in Gaborone." In the other case, marriage was never considered an option and her brother David raised the matter straight away with the Magistrate. In both cases the parties met in Gaborone, where the women were working, and the fathers of their children came from Francistown (in the eastern part of Botswana), which is a great distance from Molepolole. In these circumstances it was much more convenient for both women—who were residing in Gaborone—to deal with the man before a Magistrate, than to mobilize the *kgotla* system, which would have required assembling numerous family members from distant parts of the country. In cases such as these, the practical difficulties inherent in using the *kgotla* system make the Magistrate's court a more viable option.[51]

But not all families are prepared to use the Magistrate's court. Makokwe's youngest son, Ramojaki, would only consider raising the issue in the Chief's *kgotla*. In negotiations over his seventeen year-old daughter Okahune's first pregnancy, he would be content with the following: "if they [the man's family] agree to pay eight head of cattle, that is sufficient." Although a claim may lie for both compensation and maintenance,[52] and both have been awarded in Molepolole on occasion,[53] most people consider this unjust and that a choice must be made.[54] In Ramojaki's case he stated that "if the man's family do not pay up we will to Kgosing." He opted for Kgosing over the Magistrate's court because "We Batswana know our procedure and are inclined to report to Kgosing even although we know we can go to the D.C.'s (Magistrate's) court.[55] It is only children who go to the D.C.'s court. In a situation like this [a first child] we would go to Kgosing."

Individuals may be predisposed towards certain attitudes through the activities and networks in which they participate, but they are not bound by them. People shop around to obtain what they want. Those linked to the peasantariat's sphere of activities often have a tendency to favor the Chief's *kgotla* over the Magistrate's court, but when this fails to provide the remedy they seek they will go elsewhere. Such forum shopping works both ways, so that a claim initially taken to the Magistrate's court may at a later stage be transferred to the Chief's *kgotla*.[56] A switch of this kind may occur when a maintenance order has been made but the father fails to pay. In this situation some women turn to the Chief's *kgotla* in the hope that it will be able to exert greater pressure on the father to make him pay.

Whatever the legal forum, women experience difficulties in utilizing the legal system. In the case of the Chief's *kgotla* in Molepolole, this difficulty arises because of the limits placed on a claim for compensation or seduction, which is only available in the case of a women's initial pregnancy. This means that women have no claim when a second or subsequent pregnancy occurs. Yet given the realities of women's lives, in which pregnancy forms a natural part of their life cycle, this results in a large number of women being denied a remedy.[57] Given the ambiguity that surrounds the negotiation of a customary marriage it is often the case that women have a number of children in a relationship which they and their families consider to be leading to a marriage. By the time it is clear that the man and his family have withdrawn from the process it is often perceived as being too late to take any action because *kgotla* officials adopt the view that if more than one child is involved the appropriate forum for seeking a remedy is the Magistrate's court.[58]

On the other hand, while women can turn to the Magistrate's court for every pregnancy, there are problems with access. In the past, access was restricted in that the court only met once a week and gave priority to criminal cases. The result was a backlog of civil, chiefly maintenance, cases. To return to Oailse Timpa, she explains that she "went to report the case to the D.C. This was in November 1977. It took a long time before they were able to call us together. Our case was heard in August 1978." She was more fortunate than others who have had to wait several years for a hearing and are still waiting. The situation has been alleviated to a degree; since 1989 there has been a resident Magistrate in Molepolole, but he recognizes that delays are still a problem. The sheer volume of work means it is hard to reserve one day a week to deal with backlogged as well as current maintenance cases.

Access is not the only problem. While parties do not need legal representation and there is no fee (except on appeal from the Chief's *kgotla*), proceedings in this forum are conducted in English, a language with which the parties may have less familiarity than their local language, Setswana. For those who are unable to follow the proceedings an interpreter is provided, but this changes the dynamics of the process because another layer is added through the medium of translation. An even more important consideration is that those unmarried women who are claiming maintenance have to meet the conditions of the Affiliation Proceedings Act 1970,[59] with which they are unfamiliar. The crucial provisions are those concerning proof of paternity and the time limit for raising an action. Under the Act, a woman's allegations of paternity must be "corroborated in some material particular by other evidence to the court's satisfaction."[60] The problem is that most parties to maintenance cases in Molepolole are not legally represented and appear alone before a Magistrate. This was certainly so in 1982 and 1984, though there were some signs of representation in 1989 when a lawyer's office opened in the village. Prior to this, parties desiring legal advice and representation had to travel to Gaborone to acquire it. But for the most part there is no such representation, and where there are no witnesses or letters to establish the nature of the liaison it is one person's word against another. Under such circumstances it is difficult to meet the requirement for corroboration.

Most Magistrates are aware of this and many of them, not just in Molepolole but in Gaborone and other large towns,[61] handle this by asking the man if he admits to having had sexual intercourse with the woman around the time she conceived. If he does they treat this as a form of corroboration, giving rise to a presumption of paternity.[62] If the man denies having had sexual intercourse at the relevant time, there is little they can do. They do not believe in pressuring the man to have a blood test. Their attitude contrasts with that of officials in the Chief's *kgotla* in Molepolole, notably the Chief Regent, Mr. Mac Sechele. He claimed he was happy to apply such pressure when a man denied paternity for a first pregnancy. I never witnessed this because the man usually admitted paternity under pressure because of the *kgotla's* predisposition to back the woman's account. In such instances, the onus is on the man to prove otherwise.

It is not only the lack of knowledge about the Magistrate's court that causes women problems. They may also be in danger of being mislead by a man so that they are unable to provide the necessary corroboration for their story. This experience is a common one among young women. It leads to a second problem, that of the time limit that women face. Under the Act they must raise an action within twelve months from the date of the child's birth, or "at any subsequent time," provided the man alleged to be the father "has within twelve months next after the birth paid money or supplied food, clothing, or other necessaries for its maintenance."[63] The latter provision exists to take into account the fact that men may promise marriage and then disappear, as well as the ambiguities surrounding the negotiation of a customary marriage. But by the time it is clear that a relationship will not become a marriage, the time limit has often expired and evidence of support may be hard to produce in court. While the 1970 Act was amended in 1977,[64] this simply raised the ceiling on amounts payable under the Act and did not alter the time limit. Aware of this, a number of Magistrates admit that they simply ignore the time limit.[65] However, if the case is then appealed to the High Court, which is unusual, it will be dismissed.[66]

The problems women experience in Molepolole are duplicated elsewhere.[67] While it is true that some parties appear to make choices and engage in forum shopping, it does seem that for many women the Magistrate's court is where they may go in desperation when all else has failed. When it comes to actual payments of awards, however, neither the Chief's *kgotla* nor the Magistrate's court in Molepolole has a good track record as effective enforcement agencies. This is in keeping with findings elsewhere.[68]

Conclusion

The detailed analysis of life histories over generations not only reiterates the crucial role that access to resources (including marriage) occupies in constructing people's lives, but also highlights the gendered world in which women and men live. Thus, women's and men's experiences of procreative relationships and marriage differ. Differing access to resources gives rise to the exercise of different forms of power, which impact upon individuals' abilities to negotiate with one another. These forms of power not only differ on the basis of gender, but also vary between members of the same sex and across generations.

Negotiations in everyday life often involve individuals making various claims on others. In the cases presented, in which women are concerned their claims on men revolve round procreation. Procreation is central to women's lives and takes place in the context of both unmarried and married partnerships, which create expectations and obligations among the individuals concerned and their families. Such expectations and obligations often give rise to claims for support, or compensation for pregnancy, or for rights to property acquired during the course of the relationship. These kinds of claims are a part of the fabric of day-to-day social life, but they also represent an aspect of legal discourse.

Women's narratives reveal the extent to which women's power to negotiate their partnerships with men depends upon their position in the social order, and the stage that they have reached in their life cycle with respect to the number of children they have. Faced with the proposition that procreation is a fact of life, women must nonetheless accommodate childbearing within socially sanctioned parameters that dictate the terms for recognition of a relationship and the types of obligation that are assigned to it wherever marriage is discounted. This creates problems for women, not only because of the ambiguity surrounding the creation of a customary marriage, but also because of the social pressure to bear children they experience. Becoming a mother not only marks the transition from girlhood to womanhood, but becoming a father gives a man an enhanced status among his peers, as proof of his masculinity (regardless of the actual status of the relationship). A woman must balance the social expectation that she will have children with an assessment of the circumstances surrounding her partnership with a man. There is no doubt that having children with a man places the relationship on a footing which may enhance its credibility, but should that relationship break down, the woman and children may find themselves left in a vulnerable position because of the difficulties they face in establishing grounds for support or compensation or in enforcing their claims. Thus, power represents a key element in the consideration of women's access to law. Understanding what underlies the basis for action, or lack of action, is critical because it presents a more accurate reflection of the role that law plays in upholding power and authority with respect to different interest groups within a community.

I have highlighted the ways in which the difficulties that women face in the legal arena are related to the difficulties that they face as social actors. In making explicit the connections between the two I intend to contribute to that feminist scholarship on law dedicated to the exploration of relationships between knowledge and power and between legal status and social context.[69] Making these links explicit is crucial to an understanding of women's experiences of law. Scholars[70] elsewhere have amply documented the ways in which the basic tenets of legal ideology that form part of a Western style of law are at odds with the gendered lives of women, which the legal system ignores. This is important because when it comes to considering the issue of legal rights and the claims made by Western-style law with respect to equality and neutrality, women cannot escape from the fact that what shapes the power and authority of women within social life also has an impact on them in the legal domain.[71]

By exploring these links, my analysis, derived from Kwena ethnography, forms part of broader, feminist debates on law. It focuses on the social processes that are central to the construction of people's lives, rather than confines itself to the study of institutional forums or the formal framework that structures the relationship between common and customary law, as, for example, a legal centralist position would do. My account of people's access to law stresses the way in which individuals form parts of networks that shape their world and channel their access to resources. Such resources include the power to negotiate and the terms of reference on which this is based.

It is at this level that discussions of narrative and of power and how it is constituted are important. It is here that life histories are essential because such histories illustrate the ways in which power and the narratives through which it is explicated operate to impose distinctive forms of discourse with regard to familial relationships and property. These inform the terms upon which parties speak and how they formulate claims in respect of one another in everyday life. Such discourses amount to law (at the very least in centralist terms) when they are located within particular institutional settings. But the power to shape such discourses is not confined to, or derived solely from, these legal settings because it is generated within a broader arena, one which carries authority beyond such settings into the operational heart of everyday life. In other words, law in fact both reflects and reinforces social processes and the boundaries to which they give rise. In this context, women as individuals find themselves at a disadvantage in pursuing claims against men because of the social differentiation of power relations between men and women.

An understanding of how differential power relations are constructed provides a vehicle for discussion of the conditions under which power and its discourses may alter or be transformed. By engaging with what is otherwise left out of account in legal centrism, namely the world of economic and social differentiation, another image of law takes shape. This is one that challenges the claims to autonomy, neutrality, and equality that are attributed to law in the centralist model. An alternative approach, such as one to which this chapter contributes, radically transforms the way in which law is perceived. For although the familiar sources, institutions, and personnel associated with a centralist model of law continue to be factored into legal narratives, the ways in which they do so can no longer sustain only one representation of law to the detriment of those social actors who engage with it.

Notes

1. See Cover, R. The Supreme Court 1982 Term, Foreword: *NOMOS* and Narrative. (1983/1984) *Harvard Law Review* 97: 4–68; French, R. (1996) Of Narrative in Law and Anthropology, *Law & Society Review* 30: 417–435; Minow, M. *Law's Stories;* West, R. (1993) *Narrative, Authority, and Law.* Ann Arbor: University of Michigan Press; Abu-Lughod, L. (1993) *Writing Women's Worlds: Bedouin Stories.* Berkeley: University of California Press; Sarat, A. and T. R. Kearns (eds.) (1995) *Law in Everyday Life.* Ann Arbor: University of Michigan Press.
2. White, L. E. Subordination, Rhetorical Survival Skills, and Sunday Shoes; Notes on the Hearing of Mrs. G. Pp. 40–58 in (1991) *At the Boundaries of Law: Feminism and Legal Theory.* (ed.) M. A. Fineman and N. S. Thomadsen, New York and London: Routledge; Minow, M. (1990) *Making All the Difference: Inclusion, Exclusion, and American* Law. Ithaca, NY and London: Cornell University Press. Bellow, G. and M. Minow (eds.) (1996) *Law Stories.* Ann Arbor: University of Michigan Press.
3. For an exposition and critique of this approach see Fitzpatrick, P. (1992) *The Mythology of Modern Law.* London and New York: Routledge; Griffiths, J. (1986) What is Legal Pluralism? *Journal of Legal Pluralism and Unofficial Law* 24: 1–55; Merry, S. E. (1988) Legal Pluralism, *Law & Society Review* 22(5): 869–901; Petersen, H. and H. Zahle (eds.) (1995) *Legal Polycentricity: Consequences of Pluralism in Law.* England and USA: Dartmouth.
4. Okin, S. M. (1989) *Justice, Gender and the Family.* New York: Basic Books; Smart, C. (1984) *The Ties That Bind: Law, Marriage and Reproduction of Patriarchal Relations.*

London and Boston: Routledge & Kegan Paul; Pateman, C. (1989) *The Disorder of Women: Democracy, Feminism and Political Theory.* Cambridge: Polity Press.

5. Some examples of this scholarship include Greenhouse, C., B. Yngvesson, and D. Engel (1994) *Law and Community in Three American Towns.* Ithaca, NY and London: Cornell University Press; Lynd, A. and S. Lynd "We Are All We've Got": Building a Retiree Movement in Youngstown, Ohio. Pp. 77–99 in (1996) *Law Stories.* (eds.) G. Bellow and M. Minow, Ann Arbor: University of Michigan Press; Engel, D. M. Law in The Domains of Everyday Life: The Construction of Community and Difference. Pp. 123–170 in (1995) *Law in Everyday Life.* (ed.) Sarat, A. and T. R. Kears, Ann Arbor: University of Michigan Press.

6. This is defined as "the customary law of [a] tribe or tribal community [within Botswana] so far as it is not incompatible with the provision of any written law or contrary to morality, humanity, or natural justice" (Customary Law [Application and Ascertainment Act] No. 51 (1969) s.2).

7. This is defined as "any law, whether written or unwritten, in force in Botswana, other than customary law" (1969 Acts. 2 see note 1).

8. There are now a number of studies on this issue that have been undertaken by the Women and Law in Southern Africa Research Project, which covers nine countries in the region. In Botswana their two major studies include (1992) *Maintenance Laws and Practices in Botswana,* National Institute of Development Research and Documentation, University of Botswana, Gaborone and (1994) *Women Marriage and Inheritance,* by U. Dow. P. Kidd, published by the renamed Women and Law in Southern Africa Research Trust.

9. The Kwena are one of the Tswana polities in Botswana, each of which is often called a "tribe." The word "Bakwena" is the plural form of Kwena. An individual member of the polity is a "Mokwena." Botswana, the name of the nation of which Kweneng is a district, means the "place" of Tswana people.

10. The results of this fieldwork are set out in detail in Griffiths, A. (1997) *In the Shadow of Marriage: Gender and Justice in an African Community.* Chicago: University of Chicago Press.

11. *1981 Population and Housing Census: Summary Statistics on Small Areas (for settlements of 500 or more People.* P. 7. Central Statistics Office, Ministry of Finance and Development Planning, Gaborone: Government Printer.

12. *1991 Population Census—Preliminary Results.* Central Statistics Office, Gaborone: Government Printer.

13. C. Kerven (ed.), *Migration in Botswana: Patterns, Causes, and Consequences (Final Report National Migration Study vol. 3)* (1982) Gaborone: Government Printer.

14. *National Development Plan 1985–1991 (NDP6),* p. 11. Ministry of Finance and Development Planning, Central Statistics Office. Gaborone: Government Printer.

15. *Children, Women and Development in Botswana: A Situational Analysis* (1989), pp. 61–62. A consultant's report compiled for the joint GOB/UNICEF Programme and Planning and Co-ordinating Committee. Gaborone: UNICEF and Ministry of Finance and Development Planning.

16. *National Development Plan 1991–1997 (NDP7),* p. 9. Ministry of Finance and Development Planning, Central Statistics Office. Gaborone: Government Printer.

17. Kocken, E. M. and G. C. Uhlenbeck (1980) *Tlokweng, A Village Near Town,* p. 59, ICA Publication no. 39. Leiden: Leiden University, Institute of Cultural and Social Studies.

18. Alexander, E. (1991) *Women and Men in Botswana: Facts and Figures,* p. 17, Ministry of Finance and Development planning, Central Statistics Office Gaborone: Government Printer.

19. *Migration in Botswana: Patterns, Causes and Consequences* (1982) *Final Report of the National Migration Study, vol. 3.* Ministry of Finance and Development Planning, Central Statistics Office. Gaborone: Government Printer.

20. See Peters, P. (1983) Gender, Developmental Cycles and Historical Process: A Critique of Recent Research on Women in Botswana, *Journal of Southern African Studies*, vol. 10(1). Pp. 100–122; Kerven, C. (1984) "Academics, Practitioners and all Kinds of Women in Development: A Reply to Peters." *Journal of Southern African Studies* 10(1): 259–268.

21. These are Kgosing, Maunatlala, Mokgalo, Ratshosa, Ntoloedibe and Borakalalo.

22. According to the 1982 list in Tribal Administration.

23. For a detailed discussion of this, see A. Griffiths (Chapter 3, note 9).

24. Parson, J. "Cattle, Class and State in Rural Botswana." *Journal of Southern African Studies* 7: 236–255.

25. Cooper, D. M. *An Overview of the Botswana Class Structure and its Articulation with the Rural Mode of Productions: Insights from Selebi-Phikwe* (Dated 1980). Cape Town: Centre for African Studies, University of Cape Town.

26. See note 12.

27. Starr, J. and J. F. Collier (1989) *History and Power in the Study of Law: New Directions in Legal Anthropology.* Ithaca, NY: Cornell University Press.

28. *Urban and Rural Female-Headed Households' Dependence on Agriculture* (1979) For National Migration Study. MJDP: Statistics Office; Rural Sociology Unit; Ministry of Agriculture. Gaborone: Government Printer; Kerven, C. (1984) Academics, Practitioners and all Kinds of Women in Development: A Reply to Peters. *Journal of Southern African Studies* 10(2): 259–268.

29. *Rural–Urban Migration of Women in Botswana. Final Fieldwork Report for National Migration Study.* Botswana, Gaborone: Government Printer; The Impact of Migration on the Roles of Women. In *Final Report National Migration Study* 3. See also note 12.

30. "The Impact of Male Labour Migration on Women in Botswana." In *African Affairs* 82 (1983): 367–388.

31. *National Development Plan (NDP7) 1991–1997.* Pp. 16–17, MFDP; Central Statistics Office. Gaborone: Government Printer.

32. See UNICEF (1989) *Children, Women and Development in Botswana: A Situational Analysis.* A Consultants Report compiled for the join GOB/UNICEF Programme and Co-ordinating Committee. See also UNICEF (1993) report as above.

33. For illuminating cases involving other segments of Botswana citizens, see Motzafi-Haller, P. (1986) "Whither the 'True Bushman': The Dynamics of Perpetual Marginality." In *Proceedings of the International Symposium on African Hunters and Gatherers*, pp. 295–328, F. Rottland and R. Vossen (eds.). Sprache und Geschichte in Afrika 7(1). Sankt Augustin: Monastry of Sankt Augustin; and Wilmsen, E. N. (1989), especially pp. 257–271 in *Land Filled with Flies: A Political Economy of the Kalahari.* Chicago: University of Chicago Press.

34. This ceremony involves parents, relatives, and friends of the man who came to the woman's *kgotla* to perform a ritual involving a public request for marriage and acceptance of this request by the women's family. For further details see A. Griffiths (note 9).

35. "Marriage and Extra-Marital Sexuality: The Dialectics of Legal Change Among the Kgatla." *Journal of African Law* 21 (1977): 97–123.

36. In Botswana, individuals may marry according to customary law or register a civil or religious marriage under the Marriage Act 1970 [Cap. 29: 09]. Under customary law, a man may marry more than one wife, but not if he is already married to a woman under the 1970 Act. Nor can he marry a woman under the 1970 Act if he is already married to another woman under customary law.

37. He is descended from the second house of Sechele I (1833–1892).

38. For further discussion, see Griffiths, A. (1998) Mediation, Gender and Justice in Botswana, *Mediation Quarterly*, no. 154, pp. 335–342.

39. See A. Griffiths (Chapters 5, 6, and 7, note 9).

40. Eva's marriage was registered and *patlo* was performed. The fact that both customary and statutory forms of marriage were undertaken is not unusual. Over one third of *kgotla* members' marriages were of this kind. Patricia married in church.

41. A national survey has documented that pregnancy during adolescence has affected more than one-quarter of the adult female population. See *Teenage Pregnancies in Botswana: How Big Is the Problem and What Are the Implications.* National Institute of Development Research and Documentation and the University of Botswana: Gaborone.

42. In 61 percent of cases, that is in sixty-six out of 109 situations that had arisen by 1984.

43. In 56 percent of cases, that is thirty-seven out of sixty-six relationships.

44. That is in six cases.

45. That is in twenty-one out of thirty-seven cases.

46. These accounted for 27 percent of cases, or eighteen out of the sixty-six meetings.

47. In fourteen out of eighteen cases.

48. In seventeen out of forty-three cases.

49. A shebeen is an informal, unlicensed business enterprise, often run from home, that sells home-made beer.

50. Local people often say they are going to the D.C. when they are going to the Magistrate's court. This is because during the period of colonial overrule and prior to independence in 1966, the District Commissioner often acted as a Magistrate for his area.

51. See Griffiths, A. "Support for Women with Dependent Children in Botswana." *Journal of Legal Pluralism and Unofficial Law* 23 (1984): 1–15.

52. This is on the basis that compensation is payable to the woman's guardian for her seduction, and thus impaired marriage prospects, whereas the claim for maintenance is focused on the child or children's need for financial support. The two claims can be seen as separate and s.13 of the Affiliation Act appears to provide that a maintenance claim is still competent even where a customary court has awarded compensation for that same child's birth. However, the effect of this section in law is still an open question according to A. Molokomme, (1991) *"Children of the Fence": The Maintenance of Extra-marital Children under Law and Practice in Botswana,* pp. 74–76. Research Report/African studies Centre 46. Leiden: African Studies Centre.

53. In the case of Leepile vs. Lejone (MO/76/81) maintenance of twenty-five *pula* a month was awarded in the Magistrate's court in the knowledge that an award for seduction had already been made by a customary court.

54. These views are in line with those expressed by the general public in response to the Select Committee's investigation into this issue in 1976, prior to the 1977 amendment to the Affiliation Proceedings Act.

55. See note 48.

56. See note 49.

57. This is not the case among neighboring Bakgatla, who have altered their customary law to allow for compensation for pregnancy regardless of the number of pregnancies involved.

58. Under customary law it is competent to raise an action for compensation relating to marriage, as distinct from compensation for pregnancy. *Kgotla* officials, however, are reluctant to draw a distinction between the two and tend to treat them all as cases involving compensation for pregnancy.

59. [Cap. 28: 02].

60. S. 6(2).

61. This was information acquired during the 1982 research period.

62. This practical approach to the problem of corroboration is one which is supported to varying degrees under Roman-Dutch law in South Africa (see S. Swart 1965 (3) SA 454 (AD) and S. v Jeggels 1962 SA 704). However, the relationship between Roman-Dutch law, and legislation such as the Affiliation Proceedings Act, representing different branches of the common law in Botswana, is a problematic one. This is because the definition of "common law" includes "any law whether written or unwritten in force in Botswana, other than customary law." See note 7. Thus, two different sources of common law, that is statutory law and Roman-Dutch case law, have to be accommodated within one legal framework to produce a coherent approach to maintenance. According to Molokomme

(1991, at 81 see note 52), this problematic relationship is further complicated by uncertainty surrounding presumption of paternity and how far it extends under Roman-Dutch law in Botswana, which seems to depend upon whether judges adopt a formalist or more liberal approach to the interpretation of case law.

63. Sections 4(a) and (b).

64. By the Affiliation Amendment Act 1977.

65. While the statute presents a time limit, under Roman-Dutch law no such limit exists and a case can be brought at any time. The High Court has resolved the conflict between the two in favor of the statutory provision in the case of Tselayabotlhe Makwati vs. Olefile Ramohago, heard at the High Court on September 16, 1982 (an unreported case transcribed by Molokomme, A. and B. Otlogile (1988) *Cases on Family Law and Succession* at p. 21. Gaborone: Department of Law, University of Botswana).

66. See Makwati vs. Ramohago, note 65. The case was dismissed by the judge on the basis that the appellant did not bring her claim within the twelve month period laid down by the statute. He held that as this provision was mandatory the appellant was barred from raising a maintenance action.

67. For a detailed account of the difficulties women face in these circumstances, which cannot be presented here, the reader may refer to the work of Molokomme (see note 52); and the Women and Law in Southern Africa Research Project, Botswana (1992) *Maintenance Laws and Practices in Botswana.*

68. See Brown, B. (1985) *Report on the Child Maintenance Laws.* Women's Affairs Unit, Gaborone: Ministry of Home Affairs; Molokomme, A., p. 231, as per note 62.

69. Smart, C. (1984) *The Ties That Bind: Law, Marriage and Reproduction of Patriarchal Relations.* London and New York: Routledge; Pateman, C. *The Sexual Contract.* Stanford, CA: Stanford University Press; Minnow, M. Societal Factors Affecting The Creation of Legal Rules for Distribution of Property at Divorce. Pp. 265–279 in (eds.) M. A. Finemann and N. S. Thomadsen (1991) *At the Boundaries of Law: Feminism and Legal Theory.* New York and London: Routledge.

70. Fineman, M. A. (1991) Introduction in *At the Boundaries of Law: Feminism and Legal Theory,* (eds.) M. A. Finemann and N. S. Thomadsen. New York and London: Routledge; Mackinnon, C. (1983) "Feminism, Marxism, Method and the State: An Agenda for Theory." *Signs* 8(2): 635–658.

71. See note 4.

PART II
REFLECTIONS ON ETHNOGRAPHY IN LAW

CHAPTER 10
A FEW THOUGHTS ON ETHNOGRAPHY, HISTORY, AND LAW

Lawrence M. Friedman

At the very beginning of this short chapter, I have to lay my cards on the table. I am not an anthropologist or an ethnographer; in fact, I have never had any training in either of these fields. This makes me an outsider, as far as this volume of essays is concerned. But of course, most ethnographers are outsiders themselves, as far as their work lives are concerned. Very few of them study their own society. They tend to go to remote places and do "field work." The classical anthropologists did their work in Africa, on various Pacific Islands, or among native peoples of the Americas. They wrote about coming of age in Samoa, or the way nonliterate people in Africa settle disputes, or "the Cheyenne Way," and the like. I have of course never done this kind of field work—or field work of any kind, except for the occasional interview. I have mostly worked with legal records, especially trial court records. I have been, however, a consumer at times of what ethnographers of law have produced. The social study of law, after all, owes a great deal to anthropologists like E. Adamson Hoebel or Max Gluckman.

There is, as you all know, a big literature within anthropology about conflict, law, dispute settlement, and related topics. And it is a rich and productive literature at that. Yet ethnography seems to be suffering some sort of crisis of consciousness at the moment. It is not the only field in crisis, of course. One senses in the academy a general skepticism about knowledge, fact, research: skepticism about what is real and what is nothing but a social construct. There is a widespread suspicion that "science" is not as scientific as people think it is. I don't imagine that astrophysicists or cell biologists worry about this point of view very much, but the social sciences are pretty vulnerable and pretty self-conscious. "Value-free" social science, we are told over and over again, is a complete myth. The researcher brings her own prejudices and sensibilities to the enterprise—always. We fool ourselves into thinking we are looking through a window and recording what we see; but in fact we are only looking in a mirror, which throws back at us our own reflection. Our race, our gender, our class—we can only see reality through these lenses.

There is a special and heightened skepticism about the scientific work of outsiders—people who do not belong. Some scholars, of an advanced disposition, feel very passionately that no outsider can really understand the experience, the feelings, the mindset, of a community—particularly a community of subordinated peoples—racial minorities, for example. Of course, this is not an entirely foolish idea. Oppressed people tend to close in on themselves; they do not like to air their dirty linen in public; they feel vulnerable, and consequently secretive. But the point is more general than that. After all, a foreigner who comes to the United States and spends a year or so here can hardly be expected to understand America the way a native can. To anybody else, America is a foreign language. To Americans, it is a mother tongue.

Americans know so many things about it almost instinctively. They have imbibed it from the womb.

On the other hand, I think this point about outsiders can be taken way too far. There are things about America that only an outsider is likely to see or to grasp. After all, natives are trapped in a web of familiarity. They take too much for granted. In classical anthropology, and in the social sciences generally, it was considered a strength, not a weakness, to look at society from the outside. The scholar, pitching her tent in the community, coming from a distant culture, could question the assumptions, could explore the presuppositions, could grasp some aspects of the culture as a whole, in ways that the local could never do or would never do. And, in the case of preliterate people, the natives were not in a position to spread the word about themselves. It was the anthropologists who brought the news about these cultures to the wider world.

One of classic ethnography's greatest contributions to our understanding of human societies was its outsider perspective: a stranger's cool, impassive, sympathetic view of life inside some community or culture. Each culture no doubt guards some secrets from the outside; each has a kind of wordless language that no stranger can ever hope to penetrate. There are, in other words, *limits,* borders, frontiers, that no ethnographer can pass. But there is far more that can and will yield to objective research. The best of the classic anthropologists were very skilled at deciphering the codes of a culture and transmitting their knowledge to the world at large.

Classic ethnography made another decisive contribution to human knowledge: the general idea of cultural relativity. Nowadays, some people dismiss the pioneer anthropologists as imperialist invaders, or even as racists—and there are anthropologists who seem to enjoy this form of self-flagellation. But these accusations strike me as extremely unfair. After all, at a time when white supremacy was supposed to be a scientific fact, Franz Boas and other cultural anthropologists insisted that this was wrong; they attacked as vigorously as they could the conventional wisdom about race. And it was the anthropologists, and almost nobody else, who proclaimed to the world that preliterate and submerged cultures were worthy of respect. They argued that the cultures they studied were not "savage" or "primitive," but close-knit, coherent ways of life. If anything, the anthropologists ran the danger of romanticizing the societies they were studying. I have always had trouble convincing myself the Barotse were really so great at law-work as Max Gluckman would have us believe, but he was the expert, after all. And it was certainly right of him to demand from us a measure of respect, understanding, and close scrutiny for the rituals and procedures of this people.

Ethnography is associated in people's minds with strange people in strange, underdeveloped places, but essentially ethnography is a technique, a way of looking at a subject. You can find your subject in the Trobriand Islands or you can find it by hanging around shopping malls. Ethnography is obviously one special kind of social science enterprise. It is a technique of considering, observing, co-living with human beings of some society, or some piece of society. It uses a microscope, not a telescope. Some social scientists spend their time analyzing enormous data sets statistically. Nobody would deny that this is valid and important. Some issues lend themselves to this approach. Yet it seems to me that almost every important study of a human phenomenon has an express or implied ethnographic element—regardless of how much hard data is available.

Suppose we are interested in studying racial segregation in housing. One useful way to study it is to analyze census tract data—enter numbers into computers, develop models and equations, and analyze what comes out. One rather simple thing we can learn from the data is how many census tracts are all white, how many all black, how many are mixed. What about these mixed census tracts? Can we call them "integrated tracts?" Not without more: The information is meaningless unless you have some idea of what the data mean in human terms—on the ground, so to speak. A census tract that is all white except for a few black servants who live in small rooms inside large houses is not "integrated" in any meaningful sense. Nor is a census tract in which two white families live with their adopted black children. Another wealthy tract might be all white except for two black surgeons. Another tract is half black and half white, but there is an invisible dividing line in the middle, which the raw figures do not capture. Another is 80 percent black and 20 percent white, but the whites are elderly people who were unable to take part in a general "white flight." Still another tract is 50 percent white, 50 percent black, and it represents a genuine experiment in interracial living.

None of this is obvious from the numbers themselves. You can make educated guesses, but you cannot really know. To know, you have to look, smell, examine, touch, feel, observe. You cannot understand what a *person* is like, as a human being, from quantitative indicators, no matter how many of them you have—height, weight, dimensions of muscles or bones, computer print-outs of heart beats, DNA analysis, brain waves, and so on. You can get a (rough) physical picture of, say somebody who is sixty-five years old, weighs 200 pounds, is male, and is six feet tall. But for understanding what makes him tick, none of the figures will do. On the other hand, qualitative methods are or can be equally defective. If you don't at least try to be systematic (when this is possible), you end up with nothing better than anecdotes and war stories, chance, freak events, or impressions that may not be true. In short: Numbers without ethnography are blind; ethnography without numbers is (often) a mirage.

Much of my work has been in the field of legal history and much of it has been done in archives. At first blush, one might think that archival work is governed by rules entirely different from the rules of ethnography. After all, nobody can interview the dead. Nobody can watch them at work. All that they have left behind is pieces of paper. In archaeology, there are fragments of material culture to dig up: bits of pottery, fragments of cloth, and so on. But archives are basically nothing but words.

And yet, I insist, the same general principles apply. The ideal way to go about the work is through a combination of the qualitative and the quantitative. I have mentioned that I have worked with court records. These are in many ways tricky to deal with—much less revealing for many purposes than diaries, letters, and private papers. Nonetheless, court files can be extremely enlightening. In one study, which I carried out together with one of my students, we examined a sample of the criminal cases of Alameda County (in northern California), in the late nineteenth and early twentieth centuries. But before we even drew the sample, I spent days simply browsing among the files, reading files at random, trying to get an ethnographic feel (if I can call it that) for the nature of the material. Only then did we feel ready to count and tabulate and analyze.

Legal records are stylized and formal. They are also, by nature, argumentative. In the typical situation, in a case file, the two sides tell completely different stories. It is up to the court, then, to determine which side has the better argument, but the judge, of course, sees the facts through the lenses of his times, not ours. The historian's task, in interpreting court records, is therefore a difficult one. We can record the raw, naked data, but just as is true in contemporary times, the numbers without some sense of context are blind and misleading.

Each type of case presents its own set of problems. It is awfully laborious to get the facts and figures—how many trials, of what sort, who won and who lost—but at least it is doable. The qualitative side has its own special problems. In criminal trials, all sorts of evidence that a lay person would consider highly germane simply cannot be brought into the record (at least not openly; lawyers are often quite skilled at finding ways to sneak the evidence in). Most of us would think it very relevant in, say, a burglary case to know that the man on trial was a known burglar and had spent years in jail for that offense. That evidence is usually kept out, however. The files definitely tell stories, but we have to learn how to read these stories critically; we have to decipher them, and this is never very easy.

At a trial, when the two sides tell different stories, it is sometimes almost impossible to know where the truth really lies. Most of us suspect that Lizzie Borden really did kill her father and stepmother, but we certainly cannot be sure—and the jury set her free, after all. What we can know, however, is what kind of arguments the two sides made. These arguments are important social indicators in their own right. They tell us what sorts of images and stories the lawyers thought might be appealing. They were, after all, trying to convince a lay jury of the justice or truth of their cause. Thus, the trial records have a very special value. They are witnesses to the rhetoric that ordinary people, the kind of people who sit on juries, were likely to find persuasive. Hence the trial records are useful, even irreplaceable, sources of knowledge about norms, attitudes, social stereotypes, information that might otherwise be very hard to get at. The Salem witchcraft trials, for example, tell us about folk beliefs in a particularly vivid, concrete way: more so than sermons or treatises. This is because trials have *consequences;* they are not exercises in theory. They express ideology, of course, but in a form that can be trusted: in the form of actual behavior.

Recently, I have been working on some aspects of the history of divorce law in the United States. One of the primary sources are the thousands of divorce files that are buried in the county courts of this country. Divorce, before 1970, took the form of an adversary proceeding. A person sued for divorce, claiming that the other party had committed some sort of offense against the marriage. This was the grounds for divorce. Each state had its own list of grounds. In New York, for example, the law essentially allowed divorce only for adultery. There were ways to get around this—in the twentieth century, it was possible, for example, to travel to Nevada, which had very loose divorce laws. But this was available only for the few people with enough time and money to make the trip. New York also handed out annulments like candy—annulments are declarations that the marriage was void from the start, that it in a legal sense it never existed. Still, the plain ordinary divorce was the most common way for New Yorkers to end a failed marriage; and to get one, you had to allege and prove adultery.

Adultery was no doubt as common in New York as it is everywhere, that is, quite common. Still, many New Yorkers who wanted a divorce did not have the luxury of a spouse caught red-handed in adultery; and many did have a guilty husband but preferred not to wash their dirty linen in public. There arose in New York (and in England, another jurisdiction with very strict divorce laws) a curious kind of fiction: imitation adultery. The wife would file suit for divorce, alleging adultery. The evidence? Photographs taken in a hotel, showing her husband and another woman, sitting on a bed, both of them partly or completely undressed. In fact, the photography was staged; the woman was hired to pose for these pictures, and she was paid for her services (which almost never included actual sex). Everybody, including the judges, knew what was going on; yet the system persisted. It is an interesting question to ask, why this dual system was allowed to continue, why it was not resolved, either by stamping out the practices, or by reforming the divorce laws (this ultimately happened, but only after 1970). The point I want to make here is that what people *say* in the divorce files has to be taken with a very large grain of salt. There are certainly aspects of the files that give us data that are objective and probably reliable: names, dates, number of children, peoples' ages. But the narrative accounts of the troubles in the marriage have to be decoded; the researcher has to be a kind of cryptographer of the language of the files.

One also cannot understand the files without knowing what the law is that the parties are subverting. This is a general rule in the social study of law. One cannot understand why burglars behave the way they do, unless one knows (at least in a general way) that there is such a thing as a law against burglary; and roughly what it says. The tax laws set the rules that define the nature and limits of tax fraud. People behave in reaction to what they know or think they know about the facts of law and social control. Even when they rebel, they have to have a concept of what they are rebelling against. When they evade, they evade with the law in mind.

For this reason, even the collusive divorce cases—even the lies people tell—are revealing. But they are written in a kind of code. To understand the code, we have to understand the society in which the cryptography was written. We have to learn why people say what they do, not only by learning what the law is, but what the norms of society are. Hence, even this historical work, and even when it takes a quantitative form, is at root ethnographic. The ghostly figure of the researcher reads the lips of the dead, watches the dead at work, examines the rituals of the dead.

Divorce is, in some ways, an extreme case—extreme in the divergence between the living law and the official rules. But some divergence between these two is universal. Historical study of law, in other words, is always necessarily interpretive, even when it is rigorously quantitative (as it often should be). And archival work, historical work, parsing of court records—all this is therefore, in a way, deeply ethnographic. It is ethnographic, however, in the classical sense, in the sense of the men and women whose patient and painstaking work, whose keen eyes and ears, created the art and science of culture and led at least to the possibility of understanding what is remote and mysterious and yet intensely human in other societies and times.

CHAPTER 11

MOVING ON—COMPREHENDING ANTHROPOLOGIES OF LAW

Laura Nader

Introduction

In 1965 I began my article on "The Anthropological Study of Law" with an assertion: "It is my belief that we are just now on the growing edge of an anthropological under- standing of law in its various manifestations." Such is still my belief. I went on to confess that "the anthropological study of law has not to date affected, in any grand way at least, the theory and methodology of the anthropological discipline..."(Nader 1965: 1). Such is still true. On the other hand, the anthropological study of law has had a good deal of impact on allied fields of law and social inquiry. "Our" terrain—the non-Western other—our approaches and methods such as participant observation, as well as what we have learned about social and cultural processes through ethnography, filtered into other disciplines. Notions of critique and comparison, culture and local knowledge, and various ideas about pluralism and perception also moved horizontally into sister disciplines. Indeed, an interest in one of our key subject matters—the disputing process—spread beyond the academic world.

From 1965 to 1999 dispute resolution became an industry that penetrated the neighborhoods, the schools, the prisons, the corporations of our country while NAFTA, GATT, and the WTO all have developed the means for dealing with inter- national disputes over resources, development, and other projects of neocolonization (Nader 1999). Other things changed as well throughout this same period. Mainstream legal thought was severely shaken, and over a thirty-year period the disdain by fellow social scientists for our interest in dispute processing changed to emulation. Other social scientists also began to examine the everyday in law life. In other words, it becomes clear that the continued interest in anthropologically rooted studies concerned with the less visible faces of law and the view from below has been broadly influential. In addition, since the 1960s, the view from below has expanded upward and outward. Witness the number of anthropologists of law with or without law degrees now affiliated with law schools, the American Bar Foundation, as well as the American Bar Association.

This chapter is primarily written in the style of an intellectual autobiography,[1] brief as it is, with the intention to indicate that questions, methods, and theory are inter- twined in the happenings of the world: an improved understanding of the impact of colonialism, the machinations of the Cold War, the competition for world resources, movements to democratize the third world by exporting or importing European and American legal education, legal codes and statutes, alongside of the globalization of tastes for consumer products and services, and renewed missionary zeal, all work to destabilize what earlier anthropologists described as "societies in equilibrium."

I first went to the field in 1957 during a quieter, slower period—at least on the surface. Today my students go to the jungles of Peru and Bolivia only to trip upon NGOs (Non-Governmental Organizations), corporate enterprises, missionaries, and treasure seekers. Times change, and so the title of this little essay "Moving On"; when research questions change, so do methodologies. The second part of the title, "Comprehending Anthropologies of Law," is a signal that in this move sideways to ally ourselves with sister disciplines working in the law and society fields, something might have been lost as to what anthropology and what ethnography are all about, so perhaps some ideas that anthropologists take for granted should be made clearer. As Malinowski put it many years ago: "An ethnographer who sets out to study only religion, or only technology, or only social organization cuts out an artificial field for inquiry, and he will be seriously handicapped in his work" (1922: 11). And so we can paraphrase, "An ethnographer who sets out to study only law. ... " His admonition is especially relevant to the ethnographic study of law (or the study of law as an anthropological document) during this contemporary period in which it is fashionable to equate ethnography with qualitative work or with "hanging out," or to understand law only in relation to its most immediate and specialized context. Let me indicate what it might take minimally to be an anthropologist of law versus a psychologist, sociologist, or law and society researcher with principally legal training, first by examples from my different fieldwork experiences, followed by further commentary.

Fieldwork Sites

My fieldwork sites have been of diverse sorts, but my first experience was among the Rincón Zapotec peoples of Oaxaca, Mexico—work supported by a Mexican government grant. There I learned to study disputing in law courts in the context of the wider social and cultural organization of two small villages. In fact, my first research topic centered on spatial organization and social control: a comparative study of two Zapotec mountain villages (Nader 1964). It was among these Zapotec peoples that I had the immersion so often connected with traditional fieldwork and the doing of ethnography—eighteen months or more between 1957 and 1969 and visits intermittent since 1969. Long-term ethnographic work allows one to get at process, and I might add, allows the ethnographer to identify with those she studies and amongst whom she lives.

At the time I was working with several assumptions about order and disorder and working more or less within a positivist model: There is a limited range of disputes for any particular society (that is, all societies do not fight about all the possible things human beings could fight about); there are a limited number of formal procedures used by human societies in the prevention and/or settlement (or avoidance) of grievances (e.g., courts, contests, ordeals, go-betweens, etc.); and finally, there will be a choice in the number and modes of settlement (negotiation, mediation, arbitration, adjudication, and so on). How people resolve or manage conflicting interests and how they remedy strife situations is a problem with which all societies have to deal, and usually they find not one but many ways to handle grievances.

There were also a number of empirical questions in this early period such as what do people fight and argue about publicly, how do societies handle disputes, and what is the outcome for the individual as well as for the society, within what groups are disputes concentrated, how do disputes at one level affect those at another, and finally, what are the manifest and latent jobs of the law and how are they related to the social structure? I envisioned a qualitative and quantitative sampling of dispute cases. The case was the focus because unlike any particular form of adjudication of class of disputes or functions, the case in some form is present in every society. Furthermore, I thought that mapping the component parts of a case (as in linguistics) would produce results that could prove useful as a springboard for comparative work. The idea of generating comparisons meant that I would have to develop concepts and ideas that were more or less transcultural; Western jurisprudential ideas would not do as categories for use in comparing "non-Western" cultures. In other words, I had thought long and hard about what it would take to carry out an ethnography of law and what the pitfalls might be as well. And there were pitfalls.

Among the Rincón Zapotec, the study of social relations and social groups took me in and out of the town courts, and the town court cases took me outside the court into the community, or into other communities, especially if the disputes were intervillage; however, I was close to daily life activities in areas of subsistence, life cycle, politics, music, kinship, fiestas, and projects of modernity. In ethnography focal concerns tend to be defined broadly. In my focus on disputing I used the extended case method that had been found useful in African work, carried out interviews, participant-observed, gathered census data, and used archival documents—anything (both quantitative and qualitative) that I could get my hands on in order to produce what we then thought of as a holistic ethnography.

Fieldwork is more than participant observation and producing an ethnography of law entails a good deal more than the collection of cases. Of course, it was partial, but I aimed for the most holistic (though partial) ethnography of law that I could produce (Nader 1990). I had excellent control of Spanish but only rudimentary Zapotec, and I made over a dozen trips back to the Rincón while in the process of thinking and writing. I made a film called *To Make the Balance* (Nader 1966), which moved my attention to styles of court procedure (Nader 1969), and in 1980 PBS (Public Broadcasting System) made a second film titled *Little Injustices,* in which we tried to contrast Zapotec complaint structures with those in the United States.

After all was said and done, what finally resulted was more than a localized ethnography. It was a thick description that theorized the harmony law model. It was a study in the political economy of legal cultures. A comprehensive theory of village law models required that for any understanding of the dominant hegemony of harmony I put to one side the possibility of yet more ethnographic research and set about the task of examining historical and comparative documents that dealt with Christian missionizing and European colonialism, subjects that were ignored in earlier decades. What I discovered was the use of the harmony legal model as a means of pacification through law, first as a requirement of conquest, then as a counter-hegemonic response to over 500 years of dealing with colonization.

What I learned about ethnography from this first instance was basic: Not all fieldwork was ethnographic, and fieldwork and ethnography are concepts that should not be used interchangeably. I understood better why anthropologists are averse to spelling out fieldwork methods with greater prior specificity. We need to prepare for the unexpected and we need to be flexible in order to do so. Our stance was not to be static or rigid. We were taught that searching for the "native's point of view," differentiating between what people say they do and what the ethnographer observes they do, and doing so in depth and with a broad angle requires a set of techniques and methods for gathering and analyzing data that is inclusive not only of "background issue," but that also includes both quantitative and qualitative divides. An ethnographer could be both positivist and interpretivist, a scientist and a humanist simultaneously. Relevant to the period was the reissue of Gregory Bateson's *Naven* (1958). Bateson argued against false paradigm oppositions, something that suited my eclectic temperament.

My next fieldsite, which was interspersed with the first, was located in Lebanon. I went to Lebanon shortly after the U.S. Marines landed there in 1958. During the summer of 1961, supported by a small grant from the University of California, I located a Shia Moslem village in South Lebanon near both the Syrian and Israeli borders, a village in which I collected oral cases of conflict using Arabic as the primary language. It was a preliminary foray, part of a more general inquiry into the contemporary state of Islamic law in rural settings.

My argument was straightforward. Given that Islamic law was originally of predominantly urban origin, I wondered whether customary law predominated over Islamic law in rural settings. Even though the fieldwork was of short duration, in two and one half months I was able to answer my original question. One is more efficient the second time around. I was able to generate a model of rural–urban networking around customary law, a law that unlike Islamic law, of course, operated across Moslem and Christian religious lines (Nader 1965). Although the future of customary law even today is not clear in the Middle East, Africa, or elsewhere (indeed the very definition of customary law is in question), this short field experience, more than my secluded stay with the Zapotec, sensitized me to the different layers of law that are present wherever anthropologists go. Actually, "layers" is not the correct word, because there is commonly an intermingling of legal practices, rather than a layering, that is continuous and ongoing everywhere in the world, including our own. Nevertheless, together these two field experiences provided data for comparison of two relatively homogenous communities—Zapotec and Shia Moslem—allowing me to understand better the connection between social organization, one characterized by dual organization and the other by cross-linkage, and institutions for conflict management.

The idea of using comparison as a method for discovery inspired the Berkeley Village Law Project (Nader 1995; Nader and Todd 1978). Over a twenty-year period graduate students from Berkeley (receiving support from a variety of funding agencies) went to over fifteen different locales to study the disputing *process*. In this project my students examined disputing processes using standards of fieldwork of long duration, still concentrating attention on the collection and analysis of dispute cases within the context of social and cultural organization in small relatively bounded communities. Our single most important findings centered around conditions under which different forms or styles of dispute management occurs. Again, context provided clues as to why styles of conflict decision making varied within each single culture, as well as among cultures. In the process it also became clear that developing states were changing anthropological views of the local level as being in any way isolated from impact of larger political and economic structures. And so we all moved on.[2]

The fieldwork that followed was a break from the usual anthropological fieldsite. I decided to explore the question: How do people in a mass society such as the United States complain about products and services and with what consequences? For the first time I began work in a country of which I was a citizen. Clearly, in a country of over 280 million people, one anthropologist could not cover the whole. The focus on remedy agents to whom one carried a complaint limited the field and allowed for numbers of fieldworkers using traditional anthropological methods to examine a number of remedy agents as they worked in response to the complainant. Much had been written about the problem of delayed access, or no access at all, to United States courts, and various remedies had been offered, some leading to the development of small claims courts, regulatory agencies, and public interest law firms. But few had asked exactly how people with complaints but no access to law handled their problems.

I began this work in the middle 1970s by reading and analyzing hundreds of consumer complaint letters. The analysis opened a whole new territory for research; Americans are probably the most prolific complaint letter writers in the world. From the letters, my students and I learned that people who felt unfairly treated and yet had no access to legal protection sought redress through a variety of "third-party intermediaries," from neighborhood consumer complaint offices, to media action lines, to department store complaint desks, to unions, to consumer action groups, to the White House Office of Consumer Affairs. The persistence and inventiveness in their pursuit of justice, even after they had seemingly exhausted all avenues, was extraordinary. Thereupon began the ethnographic profiling of numbers of these cases as well the organizations to which they were taken for hearing. The extended case histories of these complaints indicate a legacy of frustration, of mistrust, of apprehension. The implications of the uneven struggle that takes place daily in a million ways between individuals and institutions, I observed, were adding up to no less than the "slow death of justice" in America.

A common theme running through the final report (Nader 1980) was one of deeply disillusioned consumers and their disillusion with government and corporations. Typically, consumers who do complain begin their search for remedy as firm believers in "the system." After enduring rebuffs and runarounds, they lose faith, often retreating into anger or apathy, but sometimes they go all the way, learn about the system, and win. Although the research was geared to discover successful instances of third-party handlers, we concluded that our society has not evolved effective systems of dealing with grievances that may be small but are critically consequential. In other words, law had not adapted to the transformations of a rural to mass industrial society. The third party intermediaries we examined included a local Better Business Bureau, a state insurance department, an automobile manufacturer, a labor union, a congressional office. In conclusion, the researchers ranked the effectiveness of intermediaries in handling grievances. The most effective were those rare organizations, such as department stores, that provide complainants with face-to-face opportunities.

This several-year study of mass consumer phenomena, supported by the Carnegie Foundation, yielded both observation and recommendation. All the ethnographers were citizens of the country in which they were studying, and the funding agency encouraged a search for successful solutions to marketplace complaints. The study itself was an early multisited research project that used ethnographic work to survey how Americans complain when they have no access to law and with what consequence (Nader 1979).

The story of a single complaint that followed from this work introduced a new ethnographic angle. While the *No Access to Law* (Nader et al. 1980) work was basically ethnographically horizontal, the follow-up examination of a father's complaint about why his son's shirt burned so quickly generated another model of work, one that followed the history of a product through regulatory agencies, manufacturers, election monies during the Nixon White House. The study of that one case documented a density of horizontal interaction among American powerholders to the exclusion of any significant vertical interaction between power-holders and in this case the victim of power transgressions. I called this multisited ethnographic approach the "vertical slice" (Nader 1980b). Notice I now use the adjective "ethnographic" rather than the noun "ethnography."

The points illustrated in these different fieldwork experiences are not only that different approaches yield new knowledge, but that different yields come together to comprise a manner of achieving understanding that is a distinct improvement on holding to one or another approach exclusively. Ethnography requires multiple approaches and mileage can pay off. I return to this point at the conclusion of my story.

After my first three or four field experiences I turned to a completely different set of experiences for insights into the meanings of something we in the West call law. I had, since my first fieldwork, moved from notions of organization, structure, and social relations to culture, specifically using the concepts of ideology and hegemony in reference to a particular type of culture. As a result of professional invitations I began to interact with the American Bar at conferences. These conferences were functionally equivalent to short-term fieldwork, although they were often brief engagements supplemented by library research and the following of legal policy debates in newspapers and journals. For example, in the 1960s there were a number of meetings between local bar groups and citizen groups in which I found myself acting as translator from one group to the other. When invited to the National Judicial College in Reno, Nevada, I had the opportunity to participant-observe judges as they visited a jail. The purpose of this volatile experiment was to allow them to discover the connections between judicial action and the people who stand before judges for sentencing.

By the mid-1970s complaints about access or no access to law and about the inefficiencies of courts themselves were so rampant that privileged solutions began to coalesce. In 1976 I was invited to the Pound Conference in St. Paul, Minnesota. As I have written elsewhere (Nader 1989), it was a rich experience and some of the pieces to the materials I had been puzzling over began to fit together. Basically, this conference was organized to discuss "a better way" to solve the access-to-law problem. It was about how to distribute legal goods in response to social movement complaints about lack of access for civil rights, environmental rights, consumer rights, women's rights, native peoples' rights, and so forth. It was about the creation of new forums and most certainly about how to deal with the legal consequences of the social movements of the 1960s. It was also a conference for beleaguered judges, a place for them to complain about their work situations and the lack of support, financial and otherwise, that they had to endure.

The potential cases generated by the 1960s social movements indicated a new set of law users who had previously had little access to the courts. At the Pound Conference these potential and real cases were referred to as the "garbage cases" and it was argued that the courts should be reserved for the important cases. That there had to be "a better way" was the theme of the then-Chief Justice Warren Burger at the conference and throughout the decade. That better way was ADR or Alternative Dispute Resolution, a way of settling these new types of cases out of court in mediation sessions or possibly in arbitration. I was struck by the language used by the chief justice and proceeded to analyze the techniques of persuasion he was using to convince the Bar and the public that his better way would relieve the American Justice system of the overload coming in as a result of social movement pressures. By the end of the conference it was clear that exhortation had won over reasoning, and rhetoric over substance.

In the years after the Pound Conference the public was immersed in alternative dispute resolution rhetoric, a rhetoric in which language followed a restricted code and formulaics that combined clusters of meaning. The pattern of assertive rhetoric was accomplished by making broad generalizations, being repetitive, invoking authority and danger, presenting values as facts. Because of his authoritative position as chief justice, Warren Burger set the tone for the language that characterized the speeches and writings of others. He warned that adversarial modes of conflict resolution were tearing the society apart. He claimed that Americans were inherently litigious, that alternative fora were more civilized, and that the cold figures of the federal courts led him to conclude that we are the most litigious people on the globe. The framework of coercive harmony began to take hold. The parallel was drawn between lawsuits and war, between arbitration and peace, danger was invoked and it was suggested that litigation was not healthy.

Burger's ADR movement could be construed as antilegal and a powerful mode of pacification. He predicted that his better way, that is the ADR movement, would take until the turn of the century to take hold. Actually, it took hold and became institutionalized with such speed that many social scientists were taken off guard. I asked myself then, was this ethnographic work I was doing—observing, participating, and writing about the Pound Conference?

The question became even more complex after years of observing the ADR movement in its many ramifications. I had come full circle from the Zapotec research which concluded that harmony ideology was part of a pacification movement that originated with Christian missionaries and colonizers. I began to observe another pacification movement using the same tactics of what I came to call "coercive harmony." Coercive harmony placed new pulls on the American justice system and attacks on the American tort system were ubiquitous. At the same time the ADR movement was going global. I then began researching ADR as a soft technology of control, looking first at international river disputes (Nader 1995) and more recently at trade phenomena (Nader 1999). While I might claim that I am participant-observing at international conferences and international trade meetings, it is clear that library research had now become the key method for documenting the dissemination of a hegemony that so quickly and so efficiently permeated a variety of institutions in the United States, and that had then moved out as part of the trend towards the Americanization of global law, which includes international law, trade agreements, and more.

Conflict has always had an ambivalent place in sociocultural anthropology. In British anthropology the Manchester school, led by Max Gluckman, argued that social conflict was functional for the maintenance of social systems, while in the United States anthropologists Bernard Siegal and Alan Beals represented conflict as a dysfunctional process produced by strains and stresses in the social system. In the early 1960s the sources of conflict as well as its functional value were conceptualized in terms of broad-gauged understandings of social organization, religion, economic interdependence, and political structures. By 1968 Ralf Dahrendorf extended the argument to point out that societies are held together not by consensus but by constraint, not by universal agreement but by the coercion of some by others. Nevertheless, by the 1970s the dialogue over conflict and harmony was shifting once again. Conflict was now portrayed as uncivilized.

Throughout all of these field experiences I was developing and refining method-ologies that suited the questions I was pursuing, but I do not think I could have accomplished my goals without that first long and intensive period of fieldwork among the Zapotec. In the 1960s and 70s and 80s I had written several articles designed to expand thinking about methods in anthropology and the social sciences more generally: "Perspectives Gained from Fieldwork" (1964), "Up the Anthropologist" (1969), "The Vertical Slice" (1980), and "Comparative Consciousness" (1994) among them. Although I value the "how" of anthropology, it was the critical "what" question that was upper most in my mind.[3] While I was not overly self-conscious about what I was doing, it became increasingly apparent to me that my essays were presenting intellectual justification for pushing beyond the invisible boundaries of what constituted the anthropology of law, and anthropology more generally, and even beyond ethnography in the more traditional sense of being tied to a single locale.

From the 1970s on, federal and state government in the United States, in concert with tribes and corporations, began to push for negotiated settlements in cases ranging from religious freedom and reparation, to water, game and fishing rights. Some years later, ADR entered the reservations via national Indian conferences, professional networks, government, and private institutions, the argument being that ADR is more compatible with "traditional" native culture and society. ADR took center stage in the struggle over nuclear waste storage on Indian lands (Nader and Ou 1998). It was elemental that barriers to thinking new about an anthropology of "law" had to be removed. And if they were not, we were not doing our job, as Malinowski noted decades ago. If the study of the harmony law model, for example, leads to a study of religious proselytizing, that is where we should go. If an understanding of complaints leads us to moral minimalisms and the construction of suburbia, so be it. If the study of ADR takes us abroad and into the political economy of disputing and trade, that is where we should be. If an understanding of why a young child's shirt burned so quickly takes us into the Nixon White House to examine election bribery, that is where we pursue the question. And if a study of nuclear waste takes us to Indian reservations that is where we go.

In other words, in ethnography, the methods are subordinate to the questions being pursued. Methods become eclectic because a loyalty to a single technique, even something like participant observation, commonly stultifies research. But also the domain of law itself needs to be recognized as artificial, as, for example, in law and society studies that are unnecessarily bounded. Indigenous systems of law that were described ethnographically as part of the indigenous culture and society became incomplete as we came to shift our entire perspective as to what constituted indigenous culture and society. In the year 2002 we include legal transplants, missionary justice, AID programs, and economic globalization as part of the local ethnographic picture. Anthropologists are correct in being uneasy about drawing boundaries.

Everyone Wants To Be an Anthropologist—But It's Not That Easy

In some ways, the research trajectory of an anthropologist usually expands after the first long period of fieldwork and what often follows, at least in my case, is not ethnography in the traditional sense, but research that moves beyond prolonged face-to-face research, but which in many ways is dependent on having had the experience of immersion requiring a long period of study and residence in a well-defined place, depending on face-to-face engagement, knowledge of the language, participation in some of the observed activities, and a greater emphasis on intensive work with people more so than survey data, for example. Our traditional research techniques have been expanded by the use of tape recorders, film, and geologic surveys for mapping but many still go and stay a long time. Some anthropologists describe ethnography as a craft, a craft that is characterized by contextual specification, on the one hand, and on the other hand, one that seriously addresses the translation problem in the final write-up of a book length monograph/writing about people as they are observed in their "natural habitat."

Over the past twenty years or so it has become fashionable to "do ethnography" as Arthur Kleinman pointed out in another context (Kleinman 1999: 76–88), however "lite" it may be. However, he adds that much of what is written "discloses not ... much serious training in this research project." He goes on to note that ethnography is "an anachronistic methodology in an era of extreme space-time compression ... it is seriously inefficient. In an era ... witnessing the hegemony of analyses based in economic, molecular biological, engineering ... ethnography is not something one picks up in a weekend retreat ... it requires systematic training in anthropology ... including mastery of ethnographic writing and social theory ... and that, too, takes time."

While it is easy to pontificate that in ethnography focal concerns tend to be broadly defined, take time, result in ethnographers identifying with those they study, and are resistant to practices of routinization because of the solitary ethos of the work, it is generally recognized today that such prescriptions do not always apply to the subject matter before us. Nelson Graburn (n.d.) is concerned with studying tourists (as versus tourism). The resulting work is fieldwork, but not ethnography, because Graburn's work is too short and deconceptualized. The difficulties of such new research circumstances result in composite descriptions of fragments; but is it ethnography?

In yet a different mode Philippe Bourgois (1999) tries to explain ethnography to public health researchers who prefer almost exclusively quantitative methodologies. They think ethnography is qualitative, Bourgois tells us. It is both quantitative and qualitative, the latter being necessary to get at process by long-term ethnographic work done to understand the daily life and cultural logic of our informants (our teachers). For the ethnographer "background issues" are frequently critical to the ethnographic thrust. Furthermore, what those we study say they do may be different from what they actually do.

Bourgois admits that within anthropology itself there have been posed false oppositions between qualitative and quantitative researchers and as well false polarization between positivist, realist, and postmodern approaches. The criticality of such polarization, I believe, becomes clearer when one examines research results in conjunction with research questions.

Discussion

It is necessary to rearticulate some basic tenets of anthropological work, as we are presently working in an era of interdisciplinary and antidisciplinary moves. As most know, disciplinary transgression is both a blessing and a curse leading to repetition, imaginative thrust, or new knowledge. At the beginning of the twenty-first century, law is of general interest to anthropology because of the critical role of law in transmitting cultural hegemonies. At the same time, interdisciplinary work may result in decontextualized and dehydrated borrowings from anthropology by researchers trained in other fields.[4] For example, the recent focus on law and everyday life is posed as a discovery when indeed what is being reaffirmed is the direction of the anthropological study of law of the past seven or eight decades. If what we wish to encourage is thick understandings of law in everyday life it might behoove us to comprehend what doing anthropology of law has meant in different historical periods. Some of the skills gained in studying local communities may transfer to new contexts.

Notes

1. See Laura Nader, "Pushing the Limits-Eclecticism on Purpose." *POLAR* 22(1) (1999): 106–109, for an even briefer attempt at recounting changes in my work on the anthropology of law.
2. Numbers of these students (Klaus Koch, June Starr, Nancy Williams, Philip Parnell, and Cathie Witty) published full-scale ethnographies.
3. See "Professional Standards and What We Study" (Nader 1976) for an argument regarding the ethical aspects of deciding what to study.

4. But see Hartog (1993) for an example of a historian doing excellent ethnographic work by means of documentary research. Hartog traces the personal history of a woman living in eighteenth-century New Hampshire as she struggles to leave an abusive husband. His story includes changing relationships to religion, law and violence, and the exercise of autonomy.

References

Bourgois, Philippe
1999   "Theory, Method and Power in Drug and HIV-Prevention Research: A Participant-Observer's Critique." In *Substance Use and Misuse* 34(14): 2153–2170.
Dahrendorf, Ralf
1968   Essays in the Theory of Society. Stanford, CA: Stanford University Press.
Gluckman, Max
1959   Custom and Conflict in Africa. Glencoe, IL: Free Press.
Graburn, Nelson
n.d.   "The Accidental Pilgrim: How Do We Know About Tourists?" Paper presented for International Academy for the Study of Tourism (IAST) Meeting, Zagreb, June 1999.
Hartog, Hendrik
1993   "Abigail Bailey's Coverture: Law in a Married Woman's Consciousness." In Austin Sarat and Thomas R. Kearns (eds.), Law in Everyday Life. Ann Arbor: University of Michigan Press.
Kleinman, Arthur
1999   "Moral Experience and Ethical Reflection: Can Ethnography Reconcile Them? A Quandary for 'The New Bioethics.'" *Daedalus* (Special Issue: *Bioethics and Beyond*), Fall.
Malinowski, Bronislaw
1922   Argonauts of the Western Pacific. New York: E. P. Dutton and Co., 1961.
Nader, Laura
1964   "Talea and Juquila: A Comparison of Zapotec Social Organization." *University of California Publications in America Archaeology and Ethnology* 48(3): 195–296.
1965a   "Choices in Legal Procedure: Shia Moslem and Mexican Zapotec." *American Anthropologist* 67(2): 394–399.
1965b   "The Anthropological Study of Law." *American Anthropologist* (Special Issue: The Ethnography of Law) 67(6) (December): 3–32.
1966   To Make the Balance (Film distribution, University of California Extension).
1969   "Styles of Court Procedure: To Make the Balance." In Law in Culture and Society. Chicago: Aldine Press, pp. 69–91.
1975   "Forums for Justice—a cross-cultural perspective." In M. Lerner, ed., *Journal of Social Issues,* The Justice Motive in Social Behavior 31(3): 151–170.
1976   "Professional Standards and What We Study." In Ethics and Anthropology: Dilemmas in Fieldwork, M. Rynkiewich and J. P. Spradley, eds., New York: John Wiley & Sons, pp. 167–181.
1979   "Disputing Without the Force of Law," in *Yale Law Journal,* vol. 88, no. 5 (Special Issue on Dispute Resolution), pp. 998–1021.
1980a   (L. Nader, ed.) No Access to Law: Alternatives to the American Judicial System. New York: Academic Press.
1980b   "The Vertical Slice: Hierarchies and Children." In Hierarchy and Society: Anthropological Perspectives on Bureaucracy, G. Britain and R. Cohen (eds.), Philadelphia, Pittsburgh: ISHI Press, pp. 31–43.
1981   Film: "Little Injustices—Laura Nader Looks at the Law." Odyssey Series PBS.
1989   "The ADR Explosion: The Implications of Rhetoric in Legal Reform." *Windsor Yearbook of Access to Justice,* University of Windsor, Ontario, pp. 269–291.

1990   Harmony Ideology: Justice and Control in a Mountain Zapotec Village. Stanford, CA: Stanford University Press.
1991   Spanish translation 1998. *Ideologia arnionica-Justicia y control en un pueblo de la montana zapoteca.* Serie: DISHA. Oaxaca, Mexico.
1995   "Civilization and Its Negotiators." In Understanding Disputes: The Politics of Law, Pat Kaplan (ed.), New York: Berg Publishers, pp. 39–63.
1999   "Pushing the Limits-Eclecticism on Purpose." *POLAR* 22(1) (May 1999).
Nader, Laura and H. Todd, Jr., ed.
1978   The Disputing Process: Law in Ten Societies, Columbia University Press.
Siegal, Bernard and Alan Beals
1960   "Conflict and Factionalist Dispute." In *Journal of the Royal Anthropological Institute* 90: 107–117.
Wellin, Christopher and Gary Alan Fine
n.d.   "Ethnography as an Occupation." Unpublished ms.

NOTES ON THE CONTRIBUTORS

JUNE STARR was one of the major figures in the ethnographic study of law and was Professor of Law at the Indiana University School of Law-Indianapolis at the time of her retirement in 2000. Between 1970 and 1995 she was a faculty member in the Department of Anthropology, SUNY-Stony Brook. She had been a Fulbright Scholar at the Indian Law Institute in New Delhi and at the Ankara Law School in Ankara, Turkey, and from 1981 to 1983 she was Professor of Sociology of Law at Erasmus University in Rotterdam. Between 1981 and 1982 she was also a Fellow at the Centre for Socio-Legal Studies, University of Oxford. Her major ethnographic and historical studies of law focused on Turkey, where she conducted research on dispute processes, social change, gender relations, and the intersections of law and power. She was the author of many important books reviews, essays, and journal articles, as well as the books *Law as Metaphor: From Islamic Courts to the Palace of Justice* (SUNY University Press, 1992), *History and Power in the Study of Law* (edited with Jane Collier, Cornell University Press, 1989), and *Dispute and Settlement in Rural Turkey: An Ethnography of Law* (E. J. Brill, 1978).

MARK GOODALE is Marjorie Shostak Lecturer in Anthropology at Emory University. His ethnographic and ethnohistorical research in Bolivia has focused on the political economy of human rights discourse, class formation among rural-legal intellectuals, and the disintegration and reconstitution of non-state legal orders. His writings have appeared in the *Journal of Legal Pluralism, Current Legal Theory, Law and Social Inquiry,* the *Journal of Rural Studies,* and *Ethnohistory,* and he contributed the "Customary Law" and "Moots" entries to *Legal Systems of the World: A Political, Social, and Cultural Encyclopedia* (ABC-CLIO, 2002). Between 1995 and 2000 he was a Fellow at the Institute for Legal Studies, University of Wisconsin Law School, and in 2000 he was a Van Calker Fellow at the Institut suisse de droit comparé, University of Lausanne. His research in Bolivia has been supported by the Organization of American States and the National Science Foundation, Law and Social Science Program.

JANE F. COLLIER is Professor of Anthropology, Emerita, at Stanford University. Her long-term ethnographic research in Chiapas, Mexico, which she initiated in the mid-1960s, focused on how growing class divisions between rich and poor contribute to changes in legal procedures that have the effect of transforming local communities. Her major publications include *Law and Social Change in Zinacantan*

(Stanford University Press, 1973), *History and Power in the Study of Law* (edited with June Starr, Cornell University Press, 1989), *Sanctioned Identities* (edited with Bill Maurer, Routledge, 1995), and *From Duty to Desire: Remaking Families in a Spanish Village* (Princeton University Press, 1997). In 2002 she was awarded the Law and Society Association's Harry Kalven Prize.

SUSAN BIBLER COUTIN teaches in the Department of Criminology, Law, and Society at the University of California, Irvine. She is the author of *Legalizing Moves: Salvadoran Immigrants' Struggle for U.S. Residency* (University of Michigan Press, 2000) and *The Culture of Protest: Religious Activism and the U.S. Sanctuary Movement* (Westview, 1993). Her current research examines the process by which Salvadoran immigrants in the United States have gone from fleeing El Salvador and being denied legal status in the United States, to being regarded as "heroes" by Salvadoran officials and gaining policy concessions from the U.S. government. This project addresses questions regarding the new meanings and forms of citizenship being created in diasporic conditions.

LAWRENCE FRIEDMAN is Marion Rice Kirkwood Professor of Law at Stanford Law School. He has researched and written extensively in the areas of law and society, trusts and estates, American legal history, and the history of criminal justice. His most recent publications include *American Law in the 20th Century* (Yale University Press, 2002), *The Horizontal Society* (Yale University Press, 1999), *American Law* (W.W. Norton & Company, 2nd edition, 1998), *The Law and Society Reader: Readings on the Social Study of Law* (edited with Stewart Macaulay, and John Stookey, W.W. Norton & Company, 1995), *Crime and Punishment in American History* (Basic Books, 1993), and *The Republic of Choice: Law, Authority and Culture* (Harvard University Press, 1990). He was president of the Law and Society Association between 1979 and 1981 and winner of its Harry Kalven Prize in 1992. He is a member of the American Academy of Arts and Sciences.

ANNE M. GRIFFITHS is Reader in Law at the University of Edinburgh and a regular Visiting Professor at the University of Texas Law School. Her major research interests include family law, alternative dispute processing, comparative law, and the anthropology of law, topics on which she has published widely. She has conducted long-term ethnographic research on family law in Botswana and is the author of *In the Shadow of Marriage: Gender and Justice in an African Community* (University of Chicago Press, 1997). She is currently engaged in a comparative ethnographic research project, funded by the Annenberg Foundation, on "The Child's Voice in Legal Proceedings" in Scotland and the United States. She is an Executive Board Member of the International Commission on Folk Law and Legal Pluralism.

SUSAN F. HIRSCH is Associate Professor and Chair of the Department of Anthropology at Wesleyan University. She is the author of numerous articles on law and society in Islamic culture, as well as the books *Pronouncing and Persevering: Gender and the Discourses of Disputing in an African Islamic Court* (University of Chicago Press, 1998) and *Contested States: Law, Hegemony, and Resistance* (edited with Mindie Lazarus-Black, Routledge, 1994). In 1997–98, she was a Fulbright Scholar at the University of Dar es Salaam, Tanzania, and in 2002 she began her tenure as a Rockefeller Humanities Fellow in Islamic Studies at the Library of Congress.

ROBERT L. KIDDER is Professor of Sociology at Temple University. He has carried out ethnographic research on lawyers and litigation participants in both India and Japan, and has studied patterns of conflict management among the Amish in Pennsylvania. He is currently conducting research on the role of Japanese lawyers in dealing with the social damages inflicted by the Great Hanshin Earthquake of 1995 in Kobe, Japan. He has contributed chapters to many law and society volumes and his writings have appeared in, among others, *Law & Society Review* and *Law and Social Inquiry.* He is the author of *Connecting Law and Society: An Introduction to Research and Theory* (Prentice-Hall, 1983).

HERBERT M. KRITZER is Professor of Political Science and Law at the University of Wisconsin-Madison, and director of the undergraduate program in Legal Studies and the Criminal Justice Certificate Program. He has conducted extensive empirical research on the American civil justice system, as well as research on other common law systems. He is the author of *The Justice Broker* (Oxford University Press, 1990), *Let's Make a Deal* (University of Wisconsin Press, 1991), and *Legal Advocates: Lawyers and Nonlawyers at Work* (University of Michigan Press, 1998), and is coauthor of *Courts, Law and Politics in Comparative Perspective* (Yale University Press, 1996). He is the editor of the multi-volume *Legal Systems of the World* (ABC-CLIO, 2002), and coeditor of the forthcoming volume *In Litigation: Do the Haves Still Come Out Ahead* (Stanford University Press, 2003). He is currently completing a book manuscript based on a major, multi-faceted study of contingency fee legal practice in the United States.

SALLY ENGLE MERRY is Professor of Anthropology at Wellesley College and co-director of the Peace and Justice Studies Program. Her work in the anthropology of law focuses on law and culture, law and colonialism, and the legal construction of race. She has written *Colonizing Hawai'i: The Cultural Power of Law* (Princeton University Press, 2000), *The Possibility of Popular Justice: A Case Study of American Community Mediation* (co-edited with Neal Milner, University of Michigan Press, 1993), *Getting Justice and Getting Even: Legal Consciousness among Working Class Americans* (University of Chicago Press, 1990), and *Urban Danger: Life in a Neighborhood of Strangers* (Temple University Press, 1981). She is currently studying the regulation of violence against women within the international human rights system, analyzing it as an example of an emergent global legal order. She is a former president of the Law and Society Association and the Association for Political and Legal Anthropology.

LAURA NADER is Professor of Anthropology at the University of California, Berkeley. Her ethnographic studies of law have covered a wide range, both theoretically and regionally. Major research projects have taken her to Mexico, the Middle

East, and throughout the United States. Earlier in her career she played a significant role in directing the emphasis in legal anthropology toward the study of disputing as a social process. Her current research and writing have focused on the impact of globalization on law, and, more broadly, the creation and meaning of central dogmas. She is the author or editor of many books, including *The Life of the Law: Anthropological Projects* (University of California Press, 2002), *Naked Science: Anthropological Inquiry into Boundaries, Power, and Knowledge* (editor, Routledge, 1996), *Harmony Ideology: Justice and Control in a Zapotec Mountain Village* (Stanford University Press, 1990), *No Access to Law: Alternatives to the American Judicial System* (Academic Press, 1980), *The Disputing Process: Law in Ten Societies* (edited with Harry F. Todd, Jr., Columbia University Press, 1978), and *Law in Culture and Society* (editor, Aldine, 1969; reprint, University of California Press, 1997). She is a member of the American Academy of Arts and Sciences and was awarded the Law and Society Association's Harry Kalven Prize in 1995.

PHILIP C. PARNELL is a social and cultural anthropologist and ethnographer. He is an Associate Professor of criminal justice at Indiana University in Bloomington with appointments in anthropology, the Center on Southeast Asia, Caribbean and Latin American Studies, and the Russian East European Institute. He has conducted ethnographic research and written about the formation of alternative legal and state systems in rural Mexico, the urban Philippines, and the United States. He is the author of *Escalating Disputes: Social Participation and Change in the Oaxacan Highlands* (University of Arizona Press, 1988), co-editor (with Stephanie Kane) of the forthcoming volume *Crime's Power: Anthropologists and the Ethnography of Crime* (Palgrave/St. Martin's Press), and is completing a book manuscript on the roles of crime, law, and disputes in urban formations of the state during the recent resurgence of Philippine democracy.

INDEX